CONSUMED

CONSUMED

FOOD FOR A FINITE PLANET

SARAH ELTON

The University of Chicago Press
CHICAGO AND LONDON

Sarah Elton is the author of *Locavore: From Farmers' Fields To Rooftop Gardens—How Canadians Are Changing the Way We Eat.*

The University of Chicago Press, Chicago 60637
The University of Chicago Press, Ltd., London

Printed in the United States of America
22 21 20 19 18 17 16 15 14 13   1 2 3 4 5

ISBN-13: 978-0-226-09362-8 (cloth)
ISBN-13: 978-0-226-09376-5 (e-book)
DOI: 10.7208/chicago/9780226093765.001.0001

Library of Congress Cataloging-in-Publication Data
Elton, Sarah, 1975–
Consumed : food for a finite planet / Sarah Elton.
pages ; cm
Includes bibliographical references and index.
ISBN 978-0-226-09362-8 (cloth : alkaline paper) — ISBN 978-0-226-09376-5 (e-book) 1. Sustainable agriculture. 2. Alternative agriculture. 3. Organic farming. 4. Agricultural ecology. 5. Food supply. I. Title.
S494.5S86 E467 2013
631.5—dc23
2013015300

♾ This paper meets the requirements of ANSI/NISO Z39.48-1992 (Permanence of Paper).

For Anisa, Nadia, and Kumail
and for all of my family

"You are a guest of nature—behave."

—FRIEDENSREICH HUNDERTWASSER

CONTENTS

INTRODUCTION

Countdown to the Future of Food

To take a look into the future, all you need do is head to the Baltimore waterfront. On a few dozen square metres of urban green space, Lewis Ziska, a USDA scientist with the Agricultural Research Service's Crop Systems and Global Change lab, found the same climatic conditions that climate models project will be the norm here on Earth sometime before the year 2050. That is, hot—about 1.5 to 2 degrees Celsius warmer than today's average temperature—and rich in carbon dioxide. Ziska used this small plot to watch in real time what effect climate change will have on the food we grow.

For his study, Ziska established two other test sites, one in the suburbs of Baltimore to simulate conditions expected in the 2020s or 2030s and another on an organic farm in Maryland to stand in for the present. (He couldn't build a laboratory to recreate these climates because funding for the kind of work he does had been cut by the federal government.) He introduced the same topsoil, with the same naturally occurring seeds, into these three landscapes and then waited to see how the different climates affected what grew. "The goal was to see what species were going to be favoured," he explained to me.

The results were unsettling. What Ziska found over the course of the six-year study was that on the steamy, carbon-dioxide-rich Baltimore waterfront, weeds grew tallest—in fact, they grew to be about two times larger than their rural counterparts. Lamb's quarters, a common leafy weed, grew to between 2.5 and 3 metres—about double the height of its cousins on that organic farm in Maryland. Ragweed also fared much better under these future-like conditions. "What we found," Ziska told me, "was that the warming winter temperatures and the high carbon dioxide were all associated with a much greater increase in the weeds. Weeds are going to be a large issue when thinking about how much food we can produce."

For those of us who don't farm, tall weeds might not sound like a big deal. But for a farmer, weeds that would tower over even an NBA player are terrifying. Their giant stalks would likely jam machinery, shade even the tallest crops, and turn farming into an all-out war between the food plants and the weeds. Farmers use herbicides today to eliminate weeds before they grow, but research shows that as carbon dioxide levels increase, these chemicals are no longer as effective. That leaves weeding by hand. "Of all the things that people do, that is the most time consuming and laborious aspect of growing food," said Ziska.

Giant weeds, resistant to herbicides, that must be pulled by hand: it's science fiction come to life. We can safely conclude that weeds stretching to three metres would compromise our food security.

"If we are concerned as a country for maintaining our national security, doesn't it make sense to remember food security?" asked Ziska. "It is an issue that is going to impact everyone's lives in the not too distant future." And by everyone, he means everyone—no matter where you live, what you choose to eat, how much you spend on groceries. In some way, every one of us will be affected. Dr. Ziska's study raises the important question of how we will feed ourselves this century.

How *will* we feed ourselves in 2050? In the next forty years, the world's population is expected to surpass nine billion. At the same time, climate change is transforming life on the planet. According to the scientists who look at these big-picture issues, in the space of about one generation, a messy combination of climate, population trends, and environmental change will profoundly affect the world as we know it. We need to figure out how to feed the world, dramatically reduce our greenhouse gas emissions, and cope with climate change.

So how *do* we best move forward? How do we ensure that everybody has enough to eat as we contend with a new climate? How do we do this without releasing even more greenhouse gases, thereby ruining the environment and further hampering the ability of future generations to feed themselves?

These pressing questions are forcing us to make a choice about how we want to tackle these problems, and a debate rages about which direction we should take. On one side of the debate is the route of sustainable food, with its organic farms and farmers' markets, seed-saving networks, short food chains, and slow food traditions. On the other side is the path of the industrial food system.

Those who believe that industrial agriculture with its worldwide economy of food will best feed the planet argue that only a global industrial food system can provide the quantity of food we need at a low enough price for people to afford. Advocates tend to conjure up a Malthusian scenario of population outstripping food supply. The image of hungry teeming masses is even used to trump the idea of sustainability—as if we as a species must make a choice between creating sustainable food systems and allowing children to die of hunger in Africa. In *Foreign Policy* magazine in 2010, Robert Paarlberg, a professor of political science at Wellesley College, slammed the sustainable food movement for what he called its elitist approach to food that excludes the poor. The subtitle summed it up: "Stop obsessing about arugula. Your 'sustainable' mantra—organic, local, and slow—is no recipe for saving the world's hungry millions." He concluded that only a globalized industrial food system can produce what we need, efficiently and cheaply, so that everyone is fed.

I disagree. I stand firmly on the other side of the debate and argue for sustainable food systems. While industrial food might provide ample quantities of cheap calories, if you want to feed people *and* protect their livelihoods given the state of the environment, the status quo doesn't cut it anymore. To feed the planet in a time of climate change, we need to build sustainable food systems. We must dramatically lessen the environmental burden of food production, encourage new economies of food that allow small-scale and family farmers to thrive in their rural communities, and nurture a food culture that connects us to the natural world on which we all depend. We must start to assemble these new, sustainable food systems immediately, because the rice, the bread, all the food we put on our plates, is at stake.

And without enough rice or bread, society starts to crack. Over the course of history, civilizations have fallen because their food systems have failed. In the Middle Ages, the Vikings disappeared from Greenland, where they had been living for several hundred years, because their farming methods eroded the topsoil and the climate changed, making it harder for them to grow food. In Central America, the Maya fled their cities, such as Guatemala's Tikal, when centuries of dry conditions, followed by drought, undermined their ability to sustain a dense urban population. The Roman Empire teetered into poverty and hunger after they overworked the soil on the plantations that supplied their busy cities. On Easter Island in the Pacific Ocean, as vividly described in Jared Diamond's book *Collapse,* the people who lived there cut down every last tree somewhere between the 1400s and the 1600s. Without trees, the Easter Islanders could no longer build the seafaring boats they had used to fish in deep waters;

they soon hunted land birds into extinction. And without trees to protect the soil from erosion, farming dwindled. They were left with few food sources other than the flesh of their human neighbours. Archaeologists have found human bones in domestic middens, their ends rounded from being boiled in a pot.

Evan Fraser is a geography professor at the University of Guelph who studies food security as it relates to climate change and economic globalization. He is also co-author of the book *Empires of Food,* which examines the role that failing food systems played in the collapse of several historical civilizations. "When we look back over ten thousand years as a species bringing food from farms to cities, the good years outweigh the bad," he explained to me. "Though history definitely reminds us that problems do emerge. History reminds us that changes do happen and happen quickly over a large scale. There are these reversals when societies do collapse very quickly.

"If you were a noble aristocrat, say, born around the year 1290, you would have been born into an affluent, confident society. You wouldn't have had any clue of what was coming unless you were paying attention to the price of wheat. Yet within two generations, 40 percent of Europe was dead. In 1315, bad rain mid-summer flattened the wheat crop and people starved. Rising demand and falling supply. The fate of society does hang in the balance."

For us, it is the year 2050 that is a bleak date in our future. That's the year when all of our environmental debts come knocking at our door, asking us to pay up. By then, the temperature of the planet will likely have risen an average of a little more than 2 degrees Celsius, with more warming at higher latitudes. That's two times more than it has risen already since 1899, around the early years of the Industrial Revolution. This warming is altering our earth's climate system. "In agriculture, the warmer it gets, the harder it will be to maintain productivity of the crops we currently consume," said David Lobell, a professor of environmental earth systems science at Stanford University who studies the impact of climate change on agriculture. There will be other effects too. Frequent droughts will reduce crop yields—as we witnessed in North America when a severe drought during the summer of 2012 limited production over a large area—and an expected increase in heavy rains will lead to floods that could destroy what we do grow. We will be forced to change the way we raise our crops, and where we grow them will be affected. A climate of extremes is bad for farming.

The results from Lewis Ziska's study in Baltimore, where weeds thrived in the heat and the high carbon dioxide levels, suggest that the way we've grown our food in the last decades—planting homogeneous rows of one crop—isn't going to fare well in a future of climate change. "Modern agriculture tends to be very large monocultures with very little genetic diversity," said Ziska. But the study demonstrates that species with the greatest genetic diversity—the ones that are the most able to produce seed and that don't rely on pollination—are the ones most able to adapt to sudden changes in climate. "Modern agriculture is the opposite," Ziska pointed out. "Is that model of agriculture good for a rapidly changing climate? Not so much."

The way we produce and consume our food in the industrial system is not only vulnerable to the effects of climate change, it is also worsening the problem. It draws heavily on fresh water (for irrigation) and fossil fuels (used to make fertilizers) and also is draining groundwater aquifers, polluting our oceans, and eroding our soils. Furthermore, the global food system is responsible for just under one-third of humanity's greenhouse gas emissions.

The predicted changes in global demographics worsen the picture. The nine billion humans that the United Nations calculates will populate the earth by 2050 is a number far greater than humanity has ever seen before. (To put it in context, in 1950, the global population was around 2.5 billion.) The majority of these people will live in cities, leaving the smallest ratio ever of farmers to eaters. As cities grow to accommodate industrial development and the rising population, as well as to cater to the housing wishes of the well-to-do who yearn for their own home and a patch of land, if things don't change we will continue to pave over the earth's best farmland, leaving ourselves with less soil on which to grow our food. At the same time, more and more people, in countries such as India and China, are embracing a Western diet, which demands more meat, hence more of nature's resources. We also depend on farmland to grow cotton and biofuels, which adds to the competition for arable land as well as the environmental burden of farming, since each industry has its own ecologically damaging practices. If we don't change our ways now, by 2050 we will struggle with the consequences of our actions.

"We are basically on the road, by the end of the century, to being like the Cretaceous period, when the dinosaurs roamed the planet, when there was no ice," earth system scientist Steven J. Davis of the University of California, Irvine, told me. By the end of the century, "we're probably talking about 6 degrees of warming. That's not enough to melt all the ice by 2100, evaporate the oceans, and drive all humans out of existence, but it is going to be unpleasant. Especially if you live in impoverished nations and at tropical latitudes."

With changing weather patterns around the world, everyone will be affected one way or another. "There will be millions of people suffering from changes in climate," said Davis. "Many will be dying from lack of food—and wars. It is plausible that the effects of climate change could cause major hits to our global economy. I don't want to sensationalize the apocalyptic scenario, but it is pretty easy to imagine that crop failures and wars in certain countries could cause the collapse of governments, and that could have an impact on the global economy."

Indeed, poor crop yields and bad weather have caused problems before. A study published in the journal *Nature* in 2011 posited that rapid changes in climate do affect farming and can be linked to conflict. Researchers at Columbia University compared incidences of civil strife during years of El Niño and La Niña, cyclic ocean-atmosphere phenomena. About every five years, the warming trends of El Niño cause droughts and torrential rains in tropical regions. The study looked at more than 200 incidents of conflict between 1950 and 2004 in 175 countries; it found that in El Niño years, there was twice the rate of conflict than in La Niña years. When turbulent weather reduces agricultural production, food prices rise and people are more likely to protest. A 2011 study by the New England Complex Systems Institute, an independent research organization, identified a point at which the price of food increases the possibility of protest. The report acknowledged that although protests are motivated by many factors, the rise in the cost of food is a likely component.

If climate already affects how we behave, what's going to happen when it becomes less predictable?

Those on the other side of the food debate say that it is precisely the industrial food system that will spare us from wars triggered by climbing food prices. Their faith in industrial food comes from the fact that it has already improved our lives. There is truth in this. There is quantitatively less poverty on earth this century than there ever has been. Far fewer of us toil in the fields as subsistence farmers, always one bad crop away from famine.

But unfortunately, industrial food has created grave problems, and now the system is sick. We have produced a world of "stuffed and starved," in the words of Raj Patel, author of the book by that name. A growing number of humans are no longer hungry—we are the stuffed. In fact, we are increasingly obese, consuming a high-fat, high-sodium, and high-sugar diet that has caused an increase in type 2 diabetes and other non-communicable diseases, as well as a childhood obesity crisis. Then there are the poor—the starved—who are chronically undernourished or who don't have enough to fill their stomachs despite the increase in yields provided by industrial food. The Food and Agriculture Organization of the UN reports that between 2010 and 2013 there were 870 million chronically undernourished people, with 98 percent of them living in developing countries. The UN's World Food Programme has found that hunger kills more people than malaria, AIDS, and tuberculosis combined. Since 2007, food prices have risen, and in 2011, the FAO's global food price index reached a historic peak. And as prices climb, more and more people go without nourishment. Food insecurity—which leaves people without access to nourishing food, or hungry, without enough to eat—is growing in the United States and Canada too. Clearly, the industrial food system has not solved our problems when it comes to feeding people. There is no salvation from the future in the status quo. If food and agriculture is the base of our civilization today, it is also the basis for its demise. Only sustainable food systems will help us to move forward.

I am not alone in this belief. Many important institutions and research bodies have reached the same conclusion that sustainable, locally centred food systems are the only way we can feed the world while reducing the damage we do to this planet. In 2009, the International Assessment of Agricultural Knowledge, Science and Technology for Development, a research project investigating the future of agriculture and funded in part by the World Bank and the United Nations, released the findings of the more than four hundred scientists, academics, and development workers it had convened from around the world. They concluded that sustainable food systems, which incorporate small-scale agriculture and traditional knowledge, should replace industrial food systems. In 2010, two more reports concurred. Olivier De Schutter, the United Nations special rapporteur on the right to food, wrote in a document he submitted to the UN Human Rights Council that policies are urgently needed to support the expansion of locally based sustainable agriculture in developing countries so that the poorest people on earth can find a better life, as well as food security, through small-scale farming. And in the United States, the National Research Council's Committee on Twenty-First Century Systems Agriculture wrote in its 2010 report that to ensure that agriculture can meet the needs of the future, farming in the United States could even require a "significant departure from the dominant systems of present-day agriculture."

But if food is the problem, it is also the solution. A new food revolution can save us. A movement is afoot that is fundamentally changing the way we produce our food and helping to provide for tomorrow. I travelled around the world to research this book, and wherever I looked, in big cities and in the smallest of villages, I found different manifestations of a similar desire to create a new way of feeding ourselves that is fair and just, that is rooted in local communities, and that doesn't have such a high environmental cost. In this book, you will read the stories of the people I met in India, China, and France, as well as in the United States and Canada. They offer us a look at what this new food order looks like.

The food system as we know it today drew on centuries of know-how, technology, and social change that allowed it to become the status quo in a mere handful of decades. If you start the count in 1945, when the Second World War ended and the industrial food system began to take shape, by 1995 food production was fully industrialized within a new worldwide agri-food economy. So if we could build it in five decades, we can replace it in about the same amount of time. To build the infrastructure and create the policies that support our ability to feed the world without damaging the environment by our deadline of 2050, we must act quickly. We must draw on the best of innovation, science, and traditional knowledge to make this new food system the new global status quo. It must be grounded in ethics, human rights, and, above all else, sustainability. Any food system we create must not hinder in any way the ability of future generations to feed themselves in a world in ecological balance.

To accomplish these goals, we must attempt to meet the following targets. The challenge for the first ten years is to stop industrial farming and make all our agricultural systems sustainable. This means an end to the widespread use of artificial pesticides and fertilizers as well as monocrops such as soybean, sugar cane, and oil palm in biodiverse regions. It also means an end to all factory-farmed meat and dairy and the transformation of the seasonless supermarket, with its endless supply of processed foods. We must localize our food markets, by closing the gap between eaters and the farmers who grow their food, to ensure that farmers can earn a living producing food and to encourage a new generation to pursue this career.

By 2030, we need to ensure that our seed supply is biodiverse. Biodiversity is the key to security; we need to preserve the genes of our food so that we can eat tomorrow. At the same time, we must support research into new varieties of crops that can survive in the new climatic conditions, and we must keep this research in the public domain so everyone around the world has access to this shared resource.

These radical changes will require a cultural shift. We need to reconnect to the planet through the food we eat. The target for 2040, then, is to embrace new food values, particularly those that appreciate seasonality and tradition and connect us to nature. To do this, we must educate the next generation as well as those who live in cities, where people are often divorced from the natural world. And this all has to happen by 2050, in time to meet the challenges of climate change, environmental damage, and population growth.

The good news is that we have already begun. From New York City to Beijing, from the Northern Hemisphere to the Southern, in rich countries and in poor, and everywhere in between, people from all walks of life are creating an alternative to the industrial food we have grown accustomed to piling into our shopping carts. This movement is about so much more than green roofs in cities or the rise of the farmers' market or the much-hyped growth in organic food's market share. This is about a rupture—a rupture with the social norms of our modern world. And the cracks this rupture is causing are already, quietly, being filled by the ingenuity of people everywhere.

An alternative is taking shape. Hundreds of thousands—probably millions—of people are devoting their lives to creating new, sustainable, and just food systems where they live. They are building on millennia worth of knowledge and practice, combining science with traditional know-how, and proving not only that the new food revolution is necessary but that it is the better way to feed us all without destroying the planet. These people offer the rest of us the opportunity to be optimistic about the possibility of change, despite the dire environmental predictions. They also invite us to join in so we can find the best solutions for feeding us all with what we have on our finite planet.

The year 2050 is just a blink away. The great unspooling of the food system must speed ahead to ensure that those who come after us can all find a place at a metaphorical global table in a world with limits.

PART ONE

TARGET 2020: SOIL

CHAPTER ONE

Table for One Billion: To See Our Future, Visit Sunny India

The road can tell you so much about where you are, especially in India. Certainly the twenty-kilometre stretch of road that connects the city of Aurangabad to Bidkin, a farming village in the traditionally agricultural state of Maharashtra, says a lot about the massive change that is under way there. On a trip from the city one morning, I passed the usual cows nosing around garbage heaps looking for food alongside wild pigs roaming free. There were the *dudhwalas,* young men with aluminum milk cans strapped to the back of their motorbikes returning to the farm from their morning deliveries; women and children balancing metal water jugs on their heads; and two boys, smartly dressed in their navy blue school uniforms, hair slicked with almond oil, walking arm in arm down the road from a vegetable seller, his wares spread on a blanket. A man soaked in a tub in front of a little shack. Another got a shave with a straight razor in a tiny roadside barbershop, lit by a single bulb hanging from a wire. A blacksmith worked a piece of metal with his hammer over an open flame. And at a cluster of stalls selling snacks and bright-coloured packs of *supari* (areca nut and betel leaf to chew), a group gathered for a breakfast of hot chai and a dish called *pava.*

Then there were the signs of the twenty-first century creeping out of the city towards the rural areas, along the road that is like a wick, drawing change to the villages and farms. Aurangabad was founded in 1610, and at least for the past fifty years, the city has been a sleepy town in an arid agricultural area where farmers grew grains such as sorghum and millet and, more recently, cash-crop cotton. That's now changing. Today Aurangabad is one of India's fastest-growing cities, in large part because manufacturing industries have moved here. In 2007, the UK-based International Institute for Environment and Development listed the city as one of the top 100 fastest-growing large urban areas in the world, and according to the Indian census, between 2001 and 2011, Aurangabad's population jumped from 2.8 to 3.6 million. About 65 percent of those people are under thirty-five years old.

When I visited, the city's first traffic lights had recently been erected in an attempt to control the growing number of motorbikes, rickshaws, trucks, and cars. Before the installation of the stoplight, a policeman wearing white gloves had directed traffic at the intersection where the road to Bidkin forks off. He now looked hopelessly at the mess of vehicles that continued past the red signal. A nearby billboard advertised a new flavour of artisanal ice cream, an appeal to a growing middle class. Then, farther along the Bidkin road, towards the outskirts of the city, fallow fields were interspersed with new housing developments of concrete multi-storey buildings. Farther still, there were signs for future construction projects. One sign, for the Sai Labh Enclave, promised "The Touch of Luxury and Comfort Living" in its bungalows, row houses, and flats. And even farther down the road to Bidkin, a man and a boy led a herd of goats through a roadside field to graze under yet another billboard offering building plots for sale. In the previous four years, the price of real estate had increased fivefold here. This is what some people call India Shining.

After that, the factories began: the steel mill with a tall chimney exuding a constant stream of black smoke, the paper mill, a plastics plant, and the Videocon campus, one of India's largest manufacturers of fridges, air conditioners, and televisions. And all along the road near the Videocon factory were little white shacks cobbled together from the Styrofoam packaging in which the electronics components likely arrived before the products were assembled in Aurangabad. Then finally: Bidkin. A bustling place, with its own collection of roadside stalls near the central bus station. This is the town where farmers come from the vicinity every Wednesday to weigh and then sell, depending on the season, their cotton or their sugarcane at the state-run wholesale market or visit the Bidkin bazaar to buy food and supplies.

But I was heading beyond Bidkin to visit a smaller village, Dhangaon. I was on my way to visit a woman named Chandrakala Bobade, an organic farmer whom I had met a few days before at a meeting in Bidkin of women farmers. Chandrakalabai—the "bai" appended as a term of respect—is a leader in her community, in a region where more than a thousand other small farmers have managed to build a resilient local food system that doesn't rely on expensive inputs. In so doing, they have improved their livelihoods and changed lives, particularly the lives of village women. And quite inadvertently, these farmers are proving that small-scale organic farming can feed a country the size of India. They are showing that their way of producing and selling food is an important part of a new sustainable food system that can feed us into the future.

The day Chandrakalabai greeted me in her home, she had pulled her long grey hair back into a braid and draped the tail of her sari over her head like a loose scarf. She wore a nose ring, a blue stone set in a gold star, and on each wrist, about a dozen matching mint-green bangles. In between her eyebrows, she had a vermilion bindi, and on her feet were *chappals,* handmade leather sandals that she wore while working in the fields. Chandrakalabai is one of the most successful farmers in the area. She is a pioneer who set an example in her village and whose work has been emulated by others. Still, she's a quiet woman who prefers to wait to speak until asked. At the meeting of extension workers hired by the Bidkin-based Institute for Integrated Rural Development, where we first met, she didn't volunteer to talk but was called upon by the others to tell her story first. When she did, she spoke matter-of-factly about what she had achieved—though her life has been far from easy. She had dealt with domestic violence. She had lived in poverty, in villages where basics such as electricity and water are an occasional luxury. And most poignantly, of the three sons she had given birth to, only one survived past babyhood.

I liked Chandrakalabai right away. I liked her thoughtfulness and how she smiled with her eyes. We got to know each other over several days, though we were unable to talk directly with each other. Chandrakalabai speaks Marathi, which I don't understand, and she didn't speak any English. We could communicate only through a translator. Despite the cultural gulf that existed between us, there were nevertheless certain details of her life she shared with me that I could relate to: the love a mother feels towards a child; the desire to work hard and to make a comfortable home for the family. She told me of her need to get away to a quiet place to do her work—she enjoys walking the few kilometres to the small shelter made from branches and straw at the edge of her fields and sit in its shade, surrounded by her crops and the sounds of birds, to do her paperwork.

The morning I arrived in her village of Dhangaon, Chandrakalabai had called a group of women to gather at her small house. It was an old stone house with a carved wooden door painted bubble-gum pink. The house was well kept. Inside were a four-poster bed, two plastic lawn chairs, a television, a machine to crush chilies into powder, and, taking up about a third of the small space, a heap of her recently harvested cotton crop that she was storing to sell as soon as the wholesale price went up. On the wall were two hand-painted murals of Hindu gods as well as old family photos. There was no kitchen in the house. Chandrakalabai had done so well as an organic farmer over the last decade that she had been able to buy another plot of land across the street to build a kitchen house, with running water and a stove. The toilet, another sign of prosperity, was outside, attached to the living quarters.

The group of us squeezed onto the bed, the chairs, and the heap of cotton—it felt soft and cozy, just as you'd imagine a pile of freshly picked cotton would—and the women began to tell me about their farms. There was Kavita, a young, smiling kindergarten teacher and cotton farmer, who soon had to excuse herself to go to class; there was Smita, who grew cotton and the yellow pigeon peas—toor dal—that are a staple, and a woman named Duarka, who grew cotton and chilies. Over the next half-hour, more women arrived, pressing into the small room to tell me of their farms and of how their lives had improved since switching to organic methods. "It's profitable because it is less cost," explained Nanda, who grew cabbage, sugar cane, bananas, and sweet lime, a delicious citrus fruit as big as a baseball with green skin and orange pulp and the taste of a mild orange.

All their stories were similar to Chandrakalabai's. About twenty years ago, she was a typical subsistence farmer, growing millet, sorghum, vegetables, and cotton using the tools of modern agriculture such as hybrid seeds, chemical pesticides, and fertilizers derived from fossil fuels—when she could afford them. She struggled. Then, in the early 1990s, Chandrakalabai heard about a way of farming that didn't rely on any of these external inputs. A non-governmental organization working in the region—the Institute for Integrated Rural Development, where she is now employed part-time as an extension worker—taught her about organic farming. Over a few years, she changed the way she farmed.

Chandrakalabai's story shows us that small farmers in the developing world can lessen their input costs and grow organically, which increases their yields. If they can then embed themselves in a local food system with a minimum of intermediaries between them and the consumer, they can earn more money and secure a better future. It's a simple story that has big implications for the rest of the world looking for an answer to how we can feed our growing population, sustainably, by 2050.

But it's hard to be optimistic about India's future when we consider climate and population. The forecasts are bleak. Demographics, environmental change, and human decisions are together creating challenges we've not yet had to face. To start, the population is spiking. Today there are more than one billion people in India; it is forecasted that by the 2020s India will surpass China as the world's most populous country. These numbers, along with a growing economy and agricultural sector, are straining the country's natural resources. India is already running out of water. People are draining the river systems and groundwater aquifers primarily to irrigate their fields. In 1950 in India, the amount of water available per person was 5400 cubic metres; by 2000, the number had dropped by more than half.[1] According to a report by the International Water Management Institute, people in India are taking two times more water from aquifers than can be replenished. As the aquifers diminish, lakes and rivers dry up, compounding the problem. And it is agriculture that uses 90 percent of the water, with domestic and industrial uses accounting for a mere 5 percent each. In fact, there is so little water to go around that in the city of Aurangabad, the municipality turns on the civic water supply in some wards for one to two hours a day, and in other parts of the city there is running water only one hour *every other day.* Farmers in the countryside don't even have that luxury. There isn't running water in most villages, and those that do have it find it in their pipes only for a few hours a day, only some days of the week. Most small farms don't have the irrigation that crops need to thrive.

At the same time, the amount of available farmland is shrinking. Not only must existing farms be divided between more and more people as the population rises, but vast swaths of agricultural land are being turned into industrial areas for manufacturing. Between 1955 and 2001, around 2.3 million hectares in India have been taken over by growing cities.[2] Around the capital, Delhi, 17 percent of the farmland disappeared to urbanization between 1992 and 2004.[3] Land for growing food is also lost when soil is contaminated by agrochemicals, sewage sludge, and municipal garbage, and when mining and other resource-extracting industries take it over. Poor land management, such as allowing livestock to overgraze, leaves the soil vulnerable to erosion by wind and rain.

Agriculture employs about half of the Indian labour force—down from two-thirds in the last decade—and the majority of Indians still live in rural areas. Yet the Indian farmer is struggling. After independence from Britain, India had a hard time feeding itself, and agriculture in the country was plagued by low productivity. The green revolution, which imported new growing technology such as pesticides and artificial fertilizers from the West, did change things, but these new tools came at a high price. Many small farmers in India spiralled into debt because they couldn't afford to pay for seeds, fertilizers, and pesticides. Chandrakalabai lives in the state of Maharashtra, known for its cotton-growing but also for its farmer suicides. Between 1997 and 2005, in Maharashtra alone, nearly twenty-nine thousand farmers killed themselves in despair, often by drinking the pesticides that had helped put them in debt. The problem isn't going away. According to a report published in the newsmagazine *India Today* during my visit, a farmer commits suicide in the country every thirty seconds.

Now add climate change to the mix. Climate modelling demonstrates that India is one of a handful of countries where agriculture will fare the worst. Because of the country's latitude near the hottest part of the earth, temperatures are predicted to rise to the point where plants such as wheat can no longer yield as much food. By 2020, according to the Indian Agricultural Research Institute, crop yields are expected to begin decreasing because of rising temperatures, though only marginally; the real drop it predicts will be felt about sixty years later. But the institute says that as soon as the 2020s, the warming conditions will affect livestock. It estimates that about 1.5 million tons of milk will be lost from the dairy industry that decade.[4] By 2080, climate change will be so severe that India may experience crop losses of between 30 and 40 percent of today's yields.[5]

This is all happening at the same time that agricultural productivity is slipping while the country's demand for food is rising. The Indian Agricultural Research Institute predicts that by the 2020s, people there will require more food than the country currently produces. That decade, economists and agricultural analysts predict that demand for domestic cereals will exceed what the country's farmers can produce by more than *20 million tons.*[6] To put this number in perspective, in 2012 in Canada—one of the world's top five wheat producers—the crop on the Prairies was around 26 million tons. To feed India in the 2020s using today's methods of industrial agriculture, the country would need to increase its agricultural land by almost the equivalent of a Canadian Prairies' worth of wheat. Either that or boost yields on the land people are already farming (when this is done sustainably it is called ecological intensification). With competing interests vying for that land, a growing population, and climate pressures, what's happening in the country has implications for the rest of the world. The challenges we face as a species are India's writ large.

The debate about the future of food, between industrial and sustainable agriculture, is taking place within circumstances that cannot be altered by anyone on either side of the divide. Any solution must take into account the environmental realities of this century. We cannot create more farmland; there is no room to expand. Instead, we need to figure out how to produce food for the future on what agricultural land we have now. Already, you can't grow food on most of the earth's surface because 70 percent of it is covered by water or ice. About 11 percent of the earth's total land area is used for crops, but we have already expanded our farms into most potentially fertile areas. The vast grasslands of North America—the breadbasket of the world—were once mixed-grass prairie ecosystems until Europeans settled in the eighteenth century. A good portion of the West African tropics was turned into farms after the 1950s. And since the 1960s and '70s, most of the world's original forests, including the breathtakingly biodiverse rainforests of South America and Indonesia, have been levelled for pulp and paper, for livestock grazing, and to grow cash crops and plants for biofuels such as soybeans in Brazil and oil palm in Indonesia. We clear more land in biodiverse places such as the Amazon at our peril. These vast forests are the lungs of our planet and help to sustain life.

A plan for feeding the future must also take water into consideration. It's not just India that's running out of the stuff. All around the world we are drawing on the earth's freshwater aquifers faster than they can replenish. Globally, too, it is farming that uses the most water, outstripping manufacturing and industry. Of the total amount of water that we draw from rivers around the world, 70 percent goes to agriculture. According to the US Geological Survey, groundwater levels have dropped significantly as a result of overuse across the country, including in Massachusetts, New York, and Florida as well as in Oregon and Washington, where irrigation, the public water supply, and industry have together dropped water levels by as much as 100 feet.[7] Climate change will further limit the amount of water that will be available to grow our food. Spring runoff from mountain glaciers supplies water for irrigation in many areas, but now that the hotter global median temperature is melting these glaciers, the water source is increasingly limited. Further, melting glaciers are no longer able to capture spring precipitation, and water will run out to sea before farmers can use it.

If industrial food remains the status quo, we will continue to lose farmland, we will continue to drain freshwater aquifers and emit tonnes and tonnes of greenhouse gases. This will not serve us in the long term. To meet the grave challenges of the future without compromising the integrity of human life on planet Earth, we must immediately begin to support the kinds of farming systems that preserve rather than consume resources, and that benefit both the world's poorest farmers as well as the rich.

As Chandrakalabai's story shows us, the transition is possible.

CHAPTER TWO

Faster, Bigger, Richer, Weaker: The Trouble with the Green Revolution

Chandrakalabai was born in 1960, thirteen years after India's independence from Britain and the partition of the country. It was a time of nation building, when political leaders were feeling optimistic about the future. Guided by the vision of India's first Prime Minister, Jawaharlal Nehru, the state invested in the economy and in infrastructure such as roads and dams. Chandrakalabai was born where most Indians lived at the time, in a small village. Her parents were farmers, growing pigeon peas, sorghum, millet, and matki—moth beans—on arid land. When Chandrakalabai was about ten years old, her family lost this land. Their village happened to be within the thirty-five thousand hectares that would be submerged when the new Jayakwadi Dam reservoir began to fill. They were paid 800 rupees for each of their seven hectares—that's about fourteen dollars a hectare in today's currency—and were told to move elsewhere. With that money they bought two hectares in another village. "It didn't feel right," Chandrakalabai remembered. "We didn't settle into a new village easily." While her family tried to adapt to their new home, Chandrakalabai went to live with her uncle.

She went to school in her uncle's town and helped to care for her young cousin. But she was bored in class and left school in grade seven. When she was fifteen, her parents arranged her marriage to a twenty-four-year-old.

That was when she arrived in Dhangaon. If it is a sleepy village today, with only a small school, a big old tree in the main square, and a few sun-beaten streets lined with small houses, back in 1975 it was even sleepier. There wasn't much to do, so Chandrakalabai went to work as a paid day-labourer in the cotton fields that were more than an hour's walk away. Her husband left to study for his bachelor of arts, and it was her mother-in-law who taught her how to farm. "I hadn't worked before, so I wasn't used to it. And it wasn't easy work," she said. A year after she was married, her first son was born. When this son was a year old, she had another boy, who later died of measles. A third son passed away when he was only seven days old. Chandrakalabai continued to pick cotton for a daily wage of six rupees—about ten cents. They tried to grow some food of their own, but the land was dry and unfertile and nothing much came up. During drought years, when there wasn't any harvest at all, she and her husband went to look for work in town, a five-kilometre walk and then a bus ride away. "My life was a really bad situation before. But after, many things fell into place."

Her introduction to sustainable farming came through a small non-governmental organization in the area. Her mother-in-law had worked with the Institute for Integrated Rural Development, which is to this day still run by the Daniels, a Christian family originally from the south of India and who live in Aurangabad, where they maintain the IIRD's main campus outside Bidkin village. The organization teaches women skills such as microfinance and organic farming, skills the women are then paid to spread to their communities. Chandrakalabai's mother-in-law urged her to apply for a position as a rural development worker there, a job that she won. Over the next decade, Chandrakalabai worked in her village, as well as others nearby, starting self-help groups where she taught women to save their money and showed them how to pool their resources to offer members credit to invest in their own small businesses.

Chandrakalabai learned a lot through the IIRD. To start, she learned how to sign her name properly. She invested in her own farming business through the self-help group she started in her village and she also learned about organic farming. Over the next five years, she made the transition to organic growing methods on her own farm. "I started with half an acre," she told me, "and I stopped putting chemical fertilizer on it. I immediately noticed the difference in the cost of production." She saved money when she didn't have to buy inputs. But there were other benefits too. "I began with vegetables, and I started to notice the difference in taste. You could even tell the difference in colour." Her experience with the transition mirrored the academic research in the field. At the beginning, her yields suffered. It took Chandrakalabai three years to see the benefits of organic agriculture when her yields improved. "The first year it seems difficult because there is less production. In the second year, I could see the difference in the soil. In the third year, I could really tell the difference. The plants coming out of the soil were green. The leaves didn't get infected. The lentils came out of the pod robust." That third year, her yields went up 30 percent.

"It feels like the darkness in my life went away," she said to me. "I have better yields—three to four lakhs' worth of yields per year from my fields! My farm has developed well. I created it from zero, so I feel really good." She smiled; her eyes glowed. "It's overwhelming—*maan bharun zatt*," she said. Which means "my heart is filled with happiness."

What is sustainable agriculture, anyway? We have lots of words for it: alternative, organic, biodynamic, and low-input, as well as permaculture and agroecology. While each of these describes a different approach to farming, all these approaches share a common philosophy. Whereas in conventional agriculture, the farmer relies on technology and human-made inputs such as chemical fertilizers and pesticides to coax food-producing plants from the soil, the sustainable farmer takes inspiration from the natural world and uses the principles of ecology as a guide. The aim is to create a balance between the many interrelated parts of an ecosystem. So sustainable farming draws on natural goods and services—soil, water, the biodiversity that exists below ground in the form of nematodes, fungi, bacteria, and algae, as well as the biodiversity in the air such as pollinators and birds—without degrading them. It tries to mimic a healthy ecosystem.

In practice this means farmers will look for natural sources of nutrients. They could plant leguminous crops, such as lentils, that have nodules in their roots that act as a home for rhizobia, beneficial soil bacteria that fix nitrogen and make it available for plant growth. In the off-season they could also sow a plant like clover, called a green manure because it too helps to build the nitrogen in the soil. Another common way of increasing soil fertility is by spreading fields with animal manure, a rich source of nitrogen. When it comes to fighting pests and disease, sustainable farmers avoid artificial chemicals and instead try to use naturally occurring pesticides, ones they can make themselves, or even natural predators. The goal, however, is to avoid disease and pests in the first place by keeping the farm ecosystem as healthy as possible. So sustainable farmers nurture the soil, adding compost and organic material and cultivating the beneficial bacteria and other organisms. The healthier an ecosystem, the more resilient it is to stress from climate, pests, and disease.

However, to do all this requires knowledge. Rather than relying only on technology, farmers must use their know-how along with their hard work and management skills to produce food sustainably. Because the farmer acts as a steward of nature, this type of agriculture offers society public goods that everybody benefits from, such as clean water, biodiversity, and soil conservation.

There is nothing novel about this approach to growing food. Farming based on ecological principles is as old as agriculture itself. Even though when people farm they exert their control over the ecosystem to create an environment that wouldn't exist in the wild without human intervention, for the better part of our agricultural history, people have done so more or less within an ecological balance. This doesn't mean all agricultural systems throughout human history have been the same or were by definition sustainable. Farmers have always innovated and dreamt up new ways of producing food—sometimes for the worse. As we saw, great civilizations have collapsed because their food systems failed. What tends to be referred to as peasant farming—small-scale, non-mechanized, subsistence agriculture—is often a life of hard work and poverty. However, there are many examples of this kind of farming system that have not only persevered over millennia but thrived.

A book that remains beloved by organic farmers around the world today, even though it was published in 1911, documents the amazing ability of humans in Asia to cultivate the same land for more than four thousand years. *Farmers of Forty Centuries* was written by an American agronomist, F. H. King, who documented his travels through China, Japan, and Korea, where he witnessed what he called a "permanent" agriculture—that is, people managing their environment in such a way that allowed them to grow food from the same soil over a long period without reducing the soil's fertility. Another hero of the global sustainable agriculture movement, Japanese farmer and author Masanobu Fukuoka, popularized the low-impact growing methods that have been practised predominantly by indigenous people around the world and created his own way of growing food that he called "do-nothing farming." His book *The One-Straw Revolution,* which Michael Pollan called "one of the founding documents of the alternative food movement," was translated into English in the early 1970s. It details Fukuoka's farming philosophy, which is based on sustainable agriculture, intuition, and, as its name implies, not doing too much work. It is an indictment of modern agricultural practice and shows how farming should exist within nature's balance. The book concludes with an anecdote about Fukuoka's journey by train to Tokyo about the time when his country was moving to industrial production. From the window, he sees "the transformation of the Japanese countryside" and writes: "The barrenness of these fields reveals the barrenness of the farmers' spirit. It challenges the responsibility of government leaders, and clearly points out the absence of a wise agricultural policy."

This transformation of the countryside that so saddened Fukuoka—a transformation that we have witnessed almost everywhere—reaches back into history. The rapid expansion of the industrial food system after the Second World War was born of centuries' worth of innovation and social change. There had already been several periods during which societies made an organized and concerted effort to improve farming. In eighteenth- and nineteenth-century Europe, people in the Netherlands, France, and Britain radically raised yields by creating a new way of farming they called the New Agriculture. The system combined livestock rearing with crop rotations. They figured out that if they grew more fodder, they could feed more animals, which would produce more manure that they could use to fertilize the fields to grow more food. This new agriculture also incorporated plants brought from the Americas, such as potatoes and corn. Around the same time in Britain, wealthy landowners fenced off the communal lands that, until then, peasants had used to grow their food and graze their livestock. In what is now known as the enclosures, they turned these commons into private property. The landless, now with no means of feeding themselves, moved to the great slums of the cities to toil in the notorious factories of the Victorian age. The new city dwellers were fed by the colonial empire that produced sugar, tea, and coffee on plantations in the subcontinent and meat, wheat, and cheese in the Americas. They worked in factories that made consumer products from agricultural products shipped back to Britain, such as jute, cotton, and rubber. This colonial structure was the foundation for the food system we know today.

Other changes continued to push us down the path to global industrial food. In the nineteenth century, the development of the railroad and the steamship facilitated the long-distance transportation of all agricultural products. In 1876, the first refrigerated steamship, *Le Frigorifique,* took meat from Argentina to France. Soon fishing boats had refrigeration on board, and ocean transportation spiked. Around the same time, science and agriculture came together to breed new varieties of seeds and better livestock. Advances in chemistry, biology, and mechanics helped to improve other aspects of agriculture. Also, at the end of the nineteenth century in Europe and North America, universal elementary school education taught peasant farmers to read and write so they could better learn the new farming techniques that were being developed. And the Industrial Revolution spread from the cotton mills and the steel factories into consumer goods of all sorts, including food. Industrial processes such as mass preservation in the form of canning and concentration of foods like orange juice led to the invention of agri-food products such as margarine, canned soups, and the precursor to Jell-O—all developed and commercialized at the end of the nineteenth century, decades before they would be adopted by the average North American family.

Up until this point, change came at a modest pace. But after the Second World War, things sped up, and quickly the food system became unlike anything we'd seen before. Even the farm was turned into something entirely different. Whereas the mixed farm had long produced a variety of crops, with livestock kept for meat, manure, and muscle to help power farm machinery, the modern farmer would instead be inspired by the Fordist ideal to focus on one thing. He could grow wheat or other grains—one field, one crop to be tended by tractor rather than horse—or keep big livestock barns for cattle, hogs, chickens, or dairy cows. The mixed farm of the past was dismissed as inefficient. This specialization, along with the new tractors, helped farms to grow bigger and to achieve economies of scale. Whereas a typical farm before the 1950s would have been about one hundred acres, today it is just over three hundred acres, with many stretching to thousands of acres.

With farms becoming bigger and mechanizing, the thought was that we didn't need as many farmers any longer. In the United States, as well as in Canada, government policy reflected the prevailing belief that too many people worked in agriculture and that it would be in the best interest of the national economy to encourage many to leave farming and take up city jobs. In the early 1960s where I live in Canada, the government passed the Agriculture and Rural Development Act to consolidate farms into larger units of production. The message from government was to go big or get out. "People to jobs" was their rallying cry. When farmers left for the city, the government bought or leased the old farms to reforest. In 1951, there were 600,000 family farms in Canada. In 2011, according to Statistics Canada, there were only a third as many.

But the change didn't stop there. Over the next decades, every aspect of farming continued to be transformed. In the 1960s, farmers were largely independent operators. These days many farmers don't work for themselves and instead sign contracts with agri-food businesses, guaranteeing to provide, by an agreed-upon date, a hog of a certain weight, or a specified number of bushels of grain, or carrots, or lettuces, or eggs. Often the company provides everything needed to produce a particular commodity, such as the seeds or the genetic stock and feed, and then sends a truck to pick up the product when the contract says so. "The farmers in North America, they are operators, they are not farmers anymore," said Tony Fuller, a professor emeritus with the School of Environmental Design and Rural Development at the University of Guelph who has spent a long career studying the effects of changing food systems. "They are the farming equivalent of a factory."

What was on the shelf at the grocery store after the Second World War reflected these big changes. The processed foods that had been invented decades earlier were ready for mass consumption. People's disposable income rose at the same time that marketing campaigns for these food products were launched, and more and more of the concoctions replaced home cooking. These foods were produced and then sold to the public through an increasingly consolidated supply chain. The large companies in manufacturing, processing, and retail grew by buying smaller business and merging with others, creating a concentrated marketplace. Food producers adopted policies of vertical integration, so they could own the means of production at every link in the food chain. And retailers grew and grew and became global megabrands such as Walmart and Carrefour. To borrow a metaphor from Michael Carolan, author of

The Real Cost of Cheap Food, this concentration has meant that the commodity chain looks like an hourglass. On the supply side of the food chain in the United States, there are the farmers, while at the consumer end there are the consumers. Separating the supply from the consumption are merely a few food processors, manufacturers, and retailers. In his book, which documents all the problems of today's industrial food system, which he criticizes for producing cheap food, he writes, "No system can survive for very long without benefiting someone . . . cheap food has its winners—big winners." Those winners are the "people, organizations and sectors of the economy that have a huge interest in maintaining the status quo." They are what make up the slim part of the hourglass.

Countless innovations, government policies, and shifts in thinking brought about the industrialization of Western agriculture. However, it was the so-called green revolution that expediently spread new ideas and agricultural technology from the West into the developing world, triggering rapid and massive social change. The period in history that we call the green revolution began around 1949 and picked up speed over the next decade. The term refers to the American-led effort to boost agricultural production around the world through research and development into new seed varieties for staple crops that grew best with modern growing techniques such as fertilizers made from fossil fuels, new pesticides like DDT, and contemporary irrigation systems. The project was undertaken by the American government working alongside two major philanthropic organizations, the Ford and the Rockefeller Foundations. While over the course of history there have been other green revolutions, no combination of ideology and technology transformed farming as much as that of the twentieth century.

The green revolution was a product of the times. Nick Cullather, a historian at Indiana University whose book *Hungry World* documents the era, describes the green revolution as Cold War strategy. In his book, he sketches the mood and the politics that gave rise to the idea that the United States could, through improvements in agriculture, eradicate Communism as well as skyrocketing birthrates and Third World poverty to boot. "There was less concern about starvation than the potential effects of Maoism and guerrilla revolutions. It was a counter-revolutionary strategy," Cullather told me. At the time, global politics were unstable. The Second World War had recently ended and the British Empire was collapsing; there was a brutal civil war in China, and across Asia, there was deep hunger as well as peasant revolts in Vietnam, Malaysia, and the Philippines. Canada and the United States, meanwhile, were thriving, particularly in agriculture. During the Second World War, North American farmers had produced so much food that they had been able to feed their own country as well as supply food to Allied troops overseas and to British civilians. The plan was to breed new varieties of staple crops, such as wheat and rice, to boost yields around the world.

A man named Norman Borlaug, a scientist with a PhD in plant pathology and genetics, is regarded as the father of the green revolution. He worked at a crop-breeding centre in Mexico, and the approach to agriculture Borlaug and his colleagues developed there would be implemented in wheat fields around the world. It was also applied to rice in the Philippines with the creation of the International Rice Research Institute, where scientists would work to breed new varieties. The goal was to create food crops that, with the help of chemical inputs, were best able to convert sunlight and water into food, and so scientists selected crop varieties that best responded to chemical fertilizers and irrigation and were well suited to a mechanical harvest. The farms they envisioned where this new kind of farming would take place were large, flat expanses of fertile soil, such as in river valleys. They bred new kinds of wheat and rice that were shorter in height and, thanks to the artificially synthesized nitrogen fertilizers, produced big seed pods, heavy with grain; the sturdy stalks of these dwarf varieties supported the large seed heads without toppling over. These new varieties, along with the irrigation techniques and the chemicals they required, quickly spread across the world. At the end of the 1960s, Borlaug's dwarf wheat was being grown on millions of acres in countries as far apart as Mexico, Pakistan, and Turkey; the new rice was coming up in the Philippines, Vietnam, and India.

Farmers around the world today produce 145 percent more than they did before the green revolution.[8] Big cash-crop-producing countries in particular, such as the United States and Canada, have benefited a lot from both the technology and policies that boosted their yields. However, whether this quantitative increase in production equals success is open to debate. Said Cullather, "Is that a success for the farmer? Is that a success for the nation? For the consumer?" It turns out that the potential to produce loads of food isn't always the answer to society's big problems. To start, much to the surprise of those who originally backed the technology, these higher yields caused social unrest. In his book, Cullather relates that in 1968, the US Agency for International Development reported that the green revolution varieties "were altering rhythms of village life in the Punjab"—the region of India where the green revolution took off—and were fuelling conflict. The wealthier farmers had the irrigation systems—or could pay to install them—that the new seed varieties required. The small and marginal farmers—which is to say the vast majority of Indians—were not able to take advantage of the new way of farming. These inequalities in the village reignited old-time tensions related to ethnic and religious differences. Some of those farmers, who couldn't afford to farm this new way, left for the cities. These pressures that pushed people off the land were repeated in every country where the green revolution unfolded.

In the words of the environmental historian J. R. McNeill, the green revolution "appealed to elites" because it helped to move labourers from the fields into the factories.[9] Large American companies such as Monsanto and Union Carbide (which would become infamous in 1984 when a chemical gas spilled from a pesticide factory in the Indian city of Bhopal, killing tens of thousands) championed the technologies, happy for the economic opportunities in countries where farmers so far hadn't bought seeds, fertilizers, or products to deal with pests. But now that farmers had to purchase inputs, these new costs cut into their revenues. The month I visited Chandrakalabai, the media was reporting that to this day, production costs of green revolution varieties and technology are so high that even in good crop years many farmers find themselves in debt.

The discrepancy between how the United States has fared in agriculture, and the weaker farming sector in other places, can be linked to the social context. American farmers are more likely to be educated. They sell their crops within a system that, for all its flaws, still delivers the food from field to consumer. But if a farmer can't read the instructions on a bag of fertilizer, she can't apply it properly. If the local moneylender charges extortionate interest rates, then she will never get ahead, even if she can secure a loan to invest in the system. While governments in countries such as India have helped to support the green revolution through such programs as fertilizer subsidies, the governments of rich countries offer industrial farmers even more financial help. And in the developing world, corruption anywhere along the supply chain compromises the whole system.

This look back at the history of our system helps us to understand why we produce our food the way we do and sheds light on the debate about the future. The reason many farmers use green revolution technology isn't because it is intrinsically better. History shows that our food system has deep roots and exists because of decisions made in the past, and because of policy and politics. It is time to evaluate the true cost of this industrial food system and take a serious look at its alternatives.

We should be supporting sustainable food systems for many reasons. To begin with, sustainable agriculture is simply better for the environment. The industrial food system does not offer us a long-term solution to feeding the planet because it destroys land and soil, consumes too much water, emits tonnes of greenhouse gases including carbon dioxide, and is utterly dependent on fossil fuels. Sustainable or regenerative agriculture, on the other hand, produces food at the same time as it provides society with ecological goods and services. Sustainable agriculture helps to conserve water, it protects the soil from erosion, and it nurtures biodiversity by creating a hospitable habitat for pollinators rather than killing them with pesticides. When the Rodale Institute, a Pennsylvania-based organic advocacy not-for-profit, compared the soil health of organic farms and conventional farms in a thirty-year-long side-by-side farming systems trial, they found that conventional agriculture produces 45 percent more greenhouse gases than organic. That's because on agro-ecological farms, more carbon is sequestered in the soil. The organic soil also held on to nitrogen longer, meaning that less leached away in the water runoff from rain or melting snow. And in times of drought, crops planted in the organic soils fared better because the soil was richer in organic matter and microbes and was therefore better able to support plants under stressful conditions.

A study published in 2009 in the *Journal of Cleaner Production* found similar results. The study compared the greenhouse gas emissions of organic wheat and conventional wheat. It concluded that conventional wheat production released more carbon dioxide into the atmosphere than organic, largely because of the emissions from the artificial synthesis of nitrogen used to make conventional fertilizer. Another study, published in 2010, found that fields farmed organically supported more butterflies than conventionally farmed fields—an indication that organic agriculture supports more biodiversity.[10] Agroecological farms are also more resilient. In October 1998, one of the most deadly hurricanes of the last two hundred years tore through Central America. Hurricane Mitch damaged billions of dollars' worth of industry and infrastructure. Eric Holt-Giménez, executive director of Food First (also known as the Institute for Food and Development Policy), compared how conventional and agroecological farms had fared in the storm. Working with information collected on 880 smallholder plots across Nicaragua, he found that farms where agroecology was practised were more resistant to storm damage than conventional farms. On average, the plots farmed agroecologically retained 40 percent more topsoil after the storm passed and lost 18 percent less arable land in landslides.[11] The way these farmers worked the land helped to protect them in the kinds of extreme weather events predicted to increase as we move towards the year 2050.

The industrial food system is praised for the efficiency of its large farms, but it is the small farms that perform better. Many studies have shown that small farms are in fact more efficient at transforming natural resources into foods. If you consider their total food output rather than yield from only one staple crop, small farms produce more food on the same amount of land than large industrial farms do. This is because, whereas large farms tend to grow crops in monocultures—one crop, one field—small-scale farmers practising alternative agriculture will typically plant many different kinds of crops on their land, even sowing a second crop in between the rows (a technique called intercropping). Small farmers also tend to incorporate livestock into general farm production, and work to produce a variety of fruits, vegetables, eggs, meat, and dairy. All this edible output will exceed the amount of food that is grown per unit of land on a large farm.[12]

The idea that a smaller farm can produce more per hectare than a big farm sounds counterintuitive, but there is a term for the phenomenon: "the inverse relationship between farm size and output"—IR for short. To better understand it, social scientists have built mathematical models that measure the output of different kinds of farms, accounting for variability in land quality and population concentration. They ask questions like, how fertile is the land? And, are farmers in one area simply working harder and growing more food because they live near a large market where they can sell more? While there is some debate in the literature, the overriding consensus is that small farms produce more gross output per hectare per year. Even in parts of the world where farmers use green revolution technology on their small plots of land, it seems that they are able to produce more than their large counterparts.[13] Various theories explain this, but the most widely recognized interpretation reasons that small farmers are free to invest more of their labour in their land, thus producing more food. Big farms, on the other hand, have to dedicate resources to managing labour and technology, which eats into their productivity.

This phenomenon is striking in Cuba, where farmers largely have been forced off fossil fuels and into organic agriculture. After the collapse of the Soviet Union, when the country lost a prime source of cheap oil and gas and the United States tightened its trade embargo, the Cuban government had to figure out how to produce food for the island without fossil fuels. Specialists in low-input agriculture from the universities taught farmers how to grow without the green revolution's help. On farms and in cities, organic agriculture took over. Today in Cuba, farmers produce 65 percent of the country's food on only 25 percent of the island's land, growing more per hectare than a commercial farm.[14] According to statistics from the Food and Agriculture Organization, Cuba imports grains, pulses, coffee, dried milk, and meat, but is able to produce a wide variety of fruits and vegetables, pork, and dairy. A lot of the organic vegetable production is done in cities, in urban gardens they call *organopónicos.* News of their success has spread wide, and there are even trips organized for tourists to see what Cubans have accomplished.

Agroecology has many advantages, and small farms can be more efficient than large farms. However, whether or not organic agriculture can produce crop yields per unit of land that are as high as what industrial farms achieve isn't as clear-cut. How much food one way of farming yields over another is an important part of the picture because the amount of food a piece of land produces determines how much land we need to farm and how much land we can leave to be wild. Proponents of industrial food often argue that it is better for the environment because its high yields allow us to spare farmland and leave it fallow. And although there are studies that show organic agriculture outperforms conventional, a lot of research demonstrates that it can't compete. The academic work that supports the idea that organic agriculture is just as productive includes the Rodale Farming Systems thirty-year trial. Not only did it find that organic agriculture delivered ecological benefits but it also recorded organic yields that matched conventional ones. During drought years, organic corn yielded 30 percent more. Yet another study, overseen by Jules Pretty at the University of Essex, looked at how a transition to agroecology improved crop production on more than twelve million farms in fifty-seven countries. It found that sustainable methods increased yields an average of 64 percent, rising to as much as 100 percent in some cases.[15] Yet other studies show that conventional farming grows more food.

In our search for the best way to feed the world, it can be hard to make sense of all the conflicting evidence. That's why Verena Seufert, a PhD student in geography, and her colleagues at McGill University decided to study the studies. They extracted data from 125 scientific studies and searched for patterns as well as for factors that could explain variations between results. They found that—it depends! In some cases, depending on soil type and farming practices, organic yields can come close to conventional yields. Mostly, however, they do fall short. Overall, organic yields were 25 percent lower. The team's results were published in the prestigious scientific journal *Nature*.

I buy organic food. I believe fundamentally in the holistic philosophy of regenerative farming—the idea that we should be cycling nutrients in our food system, striving for ecological balance. I also believe that we should design our food system in whatever way minimizes its environmental burden. So I wasn't sure what to make of Seufert's findings. But she explained to me that although the study isn't a slam dunk for the organics movement, it doesn't prove that industrial agriculture is the answer either.

"Our study shows there is not a yes or no answer to this question," she told me. "We have to look at the context. It depends on how you manage the system. It depends on where you grow crops. It depends on what crops you grow." For example, it is harder to match the yields of organic cereal crops to conventional ones, but organic legumes such as the pigeon peas Chandrakalabai grows, as well as perennials like her sweet lime trees, fare well under organic agriculture. To figure out how to apply this information to the design of the most sustainable food system, we have to understand what explains these differences. And when thinking about yields, one element is key.

Every living thing on the planet, including humans, requires nitrogen to live and to grow. When conventional farmers fertilize their fields with synthetic materials, they are mostly adding ammonia, which is a compound of nitrogen and hydrogen. While nitrogen makes up 80 percent of the earth's atmosphere, its molecule has a triple chemical bond that very few organisms can break. In nature, only several dozen genera of soil bacteria—such as rhizobia—have evolved with the ability to fix nitrogen. That is, they can break the chemical bonds in the molecule and make nitrogen available to plants so they can use it to fuel growth. Some kinds of plants, such as legumes (lentils, chickpeas, beans), as well as what are called green manures (rye and alfalfa), have tiny hairs on their roots that can sense the presence of nitrogen-fixing bacteria in the soil. The bacteria, in return, can sense the root hairs, and they and the plant send molecular signals back and forth in an attempt to see whether they can have a mutually beneficial relationship. If all goes well, the rhizobia make their home in the nodules that the plant forms on its roots. The bacteria are happy because they have a home, and the plant secures its own nitrogen supply. "It's really beautiful," said Manish Raizada, a professor in the Department of Plant Agriculture at the University of Guelph who studies ways to reduce synthetic fertilizers. "If you dig up the roots of these plants like soybeans and clover, they look like tiny beads."

Plants require nitrogen. (Raizada is developing a low-cost nitrogen test for farmers in the developing world to enable them to test their soil fertility easily.) Nitrogen is responsible for that deep green of a plant's leaf as well as the size of a grain and its protein level.

Nitrogen is also a building block in DNA and amino acids. It's important for humans too. "When we digest food, we take the nitrogen, and the enzymes in our bodies can rebuild amino acids out of it," said Raizada. "Life depends on this."

Farmers who don't use artificially synthesized nitrogen harness the power of rhizobia by planting nitrogen-fixing crops. They rotate between leguminous plants and grains, a system devised millennia ago, long before we understood biofixation. Archaeological evidence from the Middle East shows that ten thousand years ago, people were planting pulses like lentils with barley and wheat. In the New World, indigenous people also practised companion planting and grew beans and corn; in sub-Saharan Africa, early farmers planted peanuts, cowpeas, and another groundnut called bambara with root crops; in India, lentils, peas, and chickpeas were paired with rice and wheat; and in China, rice, millet, and wheat were planted with nitrogen-rich soybeans, peanuts, and peas.

Farmers enriched their fields with nitrogen too by spreading manure (not just animal but also human manure, or "nightsoil") and other organic waste such as sludge from a riverbank. They also burned crop residues, a system known as swidden or shifting cultivation. Vaclav Smil is the author of the book *Energy: Myths and Realities* and a world-renowned researcher who has spent a career studying energy, the environment, food production, and population as a professor in the Faculty of the Environment at the University of Manitoba. He says it was the limited availability of nitrogen to farmers that for much of history kept yields below a certain threshold. Then at the beginning of the twentieth century, scientists broke the barriers of nature and discovered how to fix nitrogen themselves. Using fossil fuels and what came to be known as the Haber-Bosch process, they were able to break the nitrogen bonds and synthesize ammonia to be used as fertilizer. According to Smil, this discovery was the most important scientific advancement that would have a direct effect on the world's population. He argues that without it, we could never have grown from the 1.6 billion humans we were in 1900 to the more than seven billion we are today. He estimates that by the mid-1990s, about 40 percent of the protein we ate was produced with the help of the Haber-Bosch process. Without synthetic fertilizer, he argues, we would lose almost half of the protein available to humans. In fact, he writes that the industrial methods of farming that depend heavily on synthetic fertilizer enable us to produce food for 2.3 billion more people than traditional practices ever could.

So the question this asks of sustainable agriculture is whether enough nitrogen is available through biofixation and manure to feed humanity's growing numbers without relying on the Haber-Bosch process to produce artificial fertilizer. Can we fertilize the soil without fossil fuels and still feed ourselves?

As we debate the merits of organic versus conventional farming, we as a society are tied up in an ideological knot. But if you take a scientific approach, there really isn't anything to argue about. If we want what's best for the planet, while at the same time producing enough food to feed all of us humans by 2050, then sustainable agriculture is the way forward. Full stop. It's the details of what this looks like that we must discuss.

Jules Pretty is at the fore of research into sustainable food systems. He is a professor of environment and society at the University of Essex, author of numerous books, and a fellow of the Society of Biology and the Royal Society of Arts in the United Kingdom. For ten years, he applied theory to practice at the British-based independent research organization the International Institute for Environment and Development. Pretty advocates for what people are calling sustainable intensification—that is, growing more food on the same amount of farmland by drawing on technological innovation and a capacity to work together. The most important consideration, he said, is the effect on the environment. "Whether or not it is a sustainable practice—that's the touchstone, really," he told me. "Pure organic is not going to feed the world. Pure high-end biotech isn't going to feed the world. It's a combination of all applications."

This is not an endorsement of industrial agriculture. Rather, it's an acknowledgement that the challenge for the future is so severe that we must assiduously use all the tools at our disposal to create a food system that can feed us all without destroying the ability of future generations to feed themselves. Sustainable intensification could mean helping farmers in nitrogen-deficient regions to use synthetic fertilizers most effectively with tools such as the simple soil-fertility test that Manish Raizada is developing. His dream is for a one-dollar instant test that allows poor farmers to check their soil on their own to see what nutrients are lacking. In North America, where we use too much fertilizer, sustainable intensification means figuring out ways to reduce and eliminate the environmental consequences of industrial agriculture and expand organic production in the many areas where this growing method can compete with and even surpass conventional production. "Maybe in Africa, under a certain climate, one system is better," said Navin Ramankutty, associate professor of geography at McGill University. "Maybe in India, another system is better. I don't think there is one global answer." But the top priority is protecting the environment—without this, we compromise global food security.

While yields are important, and deciding what kind of fertilizer to use has an effect on greenhouse gas emissions, there are other ways to improve the food system. Most significantly, if we wasted less of the food we produce, then we would be much better off. Just under

40 percent of the food that is grown in North America is wasted somewhere along the food chain, from farmers' fields to the consumers' kitchens. As well, as much as 70 percent of the grains produced in the United States are used to feed livestock rather than humans, and we have dedicated millions of hectares of farmland to produce biofuels rather than food. According to the European Biodiesel Board, three million hectares of farmland on the continent alone are used to produce biofuels.

And just because studies today say organic farming often doesn't produce as much food as conventional doesn't mean we can't close this yield gap in the future. Most of the billions of dollars spent on agricultural research has been put towards conventional farming. The fact that organic can compete at all against this technology shows just how much promise it does hold. In the name of sustainability, we need to direct our efforts towards improving organic methods—agroecology—and all the many ways we can ease the environmental burden of producing food.

This will require transforming not just farming itself but also how we think about food, how we sell it, and how we are involved in producing it. For sustainable farming to thrive by the 2020s, there needs to be a market for the food these agricultural systems produce. And this market must enable the farmers to earn a living while providing consumers with ready and affordable access to their food. To complement this transition to sustainable farming, we need to remake the way food is bought and sold today.

So what does a sustainable food economy look like?

CHAPTER THREE

The Money Knot: Food Prices, Profits, and the New Global Food Trade

In Bidkin village, Wednesday is bazaar day, the day when merchants and farmers and anyone with wares to sell gather in a dusty open space and hope for many customers. The afternoon I visited, the vendors draped tarps over poles or strapped umbrellas to sticks and positioned them over their merchandise in search of shade from the hot sun. There were freshly picked vegetables: husked corn, green onions, potatoes, fresh fenugreek, okra, and ginger; coconut and paan leaves and overripe papaya and dried fish. The place was rich with colour. The deep ochre of ground turmeric, spilled on a piece of cardboard and moulded into a pyramid, the bright red of chili powder, the soft brown of cumin. A light breeze was blowing and there was enough spice dust in the air to catch in my throat and cause me to cough. There was also the strong scent of coriander and burning incense. The merchants selling jaggery, a raw sugar that's boiled to a thick mass and sold in blocks, sat grouped together in one area, while those who offered fried treats such as savoury pakoras and sweet jelabis formed their own cluster. A man shouted out to attract customers: "This is good! Good price! Good quality!"

Then there were the tools for sale. A sickle, an aluminum pot to collect water. There were bangles and slippers and tea cups and sturdy shopping bags. In the morning there had been a cattle market for those who wanted to buy livestock. Across the road was the government station where farmers and middlemen brought their cotton or sugarcane to sell during harvest season. There was an enormous heap of white cotton in the yard; some young men who looked like they were having fun bounced on the pile while a lone farmer, an older-looking thin man with a cloth wound around his head, tried to sell his crop. But soon the farmer left, unsuccessful, with his cotton bundle tied to his back. The government buys cotton only by the quintal—one quintal is 100 kilos—and the farmer didn't have enough.

According to my guide for the afternoon, Alka Najan, a woman who has lived in the district her whole life, the bazaar is the best place to shop. She explained that the prices for food at the bazaar were consistently the lowest. If you missed coming on Wednesday, she said, you could pick up vegetables from a seller by the bus stand in town, but prices there were always more expensive. The food at the Bidkin market was cheapest because there was so much competition between the dozens of vendors. When a row of merchants all sell more or less the same product, the only way they can differentiate themselves and attract shoppers is by lowering their price. For the customer, this means the food is inexpensive. For the sellers, it means margins are small.

All over India, in cities as well as towns, there are markets just like Bidkin. These, and the *kirana* stores—small family-owned shops that line so many of the country's roads—are where most Indians buy their food, and more. Much of India's more than $400 billion in annual retail sales takes place in this informal sector that employs forty million people. For now. In 2012, the government of India changed the rules overseeing direct foreign ownership that had kept the world's biggest supermarket chains out of the country. Soon after, Walmart announced its plans to open stores there. In Aurangabad, I was told that a new building on the road to Bidkin, painted an unappealing brown, was going to be a Walmart. The sign out front read only, "Best price—opening soon."

Not that this pleased many Indians. After news broke of Walmart's pending arrival, protests erupted in cities such as Bangalore and Calcutta. The crowds blocked railroad tracks and burned effigies of the prime minister. The Confederation of All India Traders told the BBC that it feared that foreign-owned megastores would ruin "the economic and social fabric of the country." The reaction wasn't only protectionist backlash. In a column in the national newspaper *The Hindu,* well-known food and agriculture watcher Devinder Sharma observed that the government's logic—that this move to make way for more supermarkets would help farmers—was baseless. Since the big food retailers in the United States haven't made farming profitable for American farmers, he asked, why would they in India?

But according to Joy Daniel, head of the IIRD in Bidkin, the local bazaar and the *kirana* shops haven't worked for small farmers like Chandrakalabai either. That's why the organization has set up an entirely new way of selling food in Maharashtra—a new way of doing business that might make it possible for farmers to earn a living growing food. It is what they call an organic bazaar and it looks like what we in America consider to be a farmers' market. Every week, a group of small organic farmers comes together at the side of a busy road in downtown Aurangabad, under the front awning of a vacant building, to display their wares and sell their produce directly to the customer. This way of doing business is a rejection of the status quo in India, but it isn't replicating the Western supermarket either. It really *is* a third way.

Certain aspects of the organic bazaar resembled the traditional market in Bidkin. When I visited, there were heaps of produce for sale—bundles of green spinach and onions, ginger, chilies, fresh turmeric, all piled on burlap sacks. A man, barefoot with a green apron tied around his waist, sliced a green squash that was even bigger than a watermelon, and weighed a piece for an older woman. Another man, with a white bandana tied around his head and selling shiny purple and green globe-shaped eggplants, as well as pumpkins, waited for customers while he rubbed the earth from his garlic. These were the tiniest heads of garlic, no bigger than a large marble; I had seen them at the other bazaar and been told they were a local variety. And just as in Bidkin, there was the smell of coriander, somehow overpowering the fumes from the incessant stream of traffic going by.

But in many other ways, it was different. At the organic bazaar, there was the chance for farmers to speak with the people who were buying their food. One goal of the market is to create an opportunity for the public to learn about organics. The logic is that once people believe that organic food has value, they will be willing to pay more for this quality produce. Chandrakalabai was helping to oversee the market that day, so she hadn't brought her produce to sell. I asked her what she says to people who ask why they should shop here, and she pretended to speak to a doubting marketgoer: "These vegetables are organic. It is good for my health and for my kids. It's tasty. It poses no danger to the body." She explained to me, "You have to fix it in their head, and once they are convinced, they will come back."

Another difference between the two bazaars: at the organic bazaar, not only do the farmers retain all the profits from their sales (after they pay a small fee for a table) but they don't undercut one another's prices either. Instead, they agree on a selling price they arrive at by adding about 20 percent to the conventional retail price, and they stick with it.

Near the edge of the road, an older-looking woman wearing a blue sari squatted in front of a heap of jujube fruit, tiny garlic bulbs, lemons, bunches of mint, and chilies. Two shoppers wearing black niqabs picked out jujubes from the pile, and the farmer weighed them on an old metal scale she raised in the air to balance, her red bangles jingling. She told me that it was Chandrakalabai and Joy Daniel of the IIRD who had told her about the organic bazaar and invited her to participate. Before, she'd sold her produce at the Bidkin bazaar, but in Aurangabad she could make more money. The premium she earned for her organic produce is significant, particularly because her farm is tiny—a mere few square metres of land. She was able to earn a living by using trellises to grow a climbing squash called durka, by planting high-earning crops such as herbs and garlic, and by harvesting lemons from her one tree.

Even though prices were higher, the organic food at the bazaar wasn't catering to elites. The woman who translated for me while I was in Maharashtra, a graduate of environmental sciences named Gauri Mirashi, who is from Aurangabad, explained that while organic food in India often caters to the wealthy, this organic bazaar did not. She noticed that there were middle-class people who shopped there, and she said the prices were comparable to those at the market where her mother shopped. I did a bit of price comparison of my own and found that okra at the organic bazaar cost less than the conventional stuff at the new supermarket in the Aurangabad mall; the organic okra's price was comparable to the asking price at the Wednesday Bidkin bazaar.

The model has been successful. About ninety farmers take turns selling through the organic bazaar during the rainy season, when vegetable production is best; from March to September, about forty farmers rotate through. In Aurangabad, a second bazaar has opened, at the local government office, and throughout the week, the farmers sell their spices, grains, and pulses in a store called Organic Link. The IIRD is also thinking of opening a stall in Bidkin. Organic bazaars have sprouted up in other Indian cities too.

Despite the success, there are still many hurdles. Rural roads are notoriously poor in India, and it can be extremely difficult for a farmer to get food to market. When you are transporting fresh vegetables such as fenugreek and okra, the time it takes to get your food to market in typically hot weather is critical. And supply chains to link farmers to their customers don't really exist yet. Nevertheless, there is a lot of potential for the future if farmers adopt this trade model and sell directly to the consumer.

The story of the organic bazaar in Aurangabad demonstrates how farmers who extricate themselves from the traditional sales model benefit by building an alternative local food economy. In so doing they preserve their autonomy, maintain control of the system, and are able to make a living growing food. Because they can make money from farming vegetables, they are less dependent on low-earning cash crops such as sugar cane and cotton. This model helps us to achieve our goals of feeding people because it keeps small farmers working the land, growing food. It also reduces the effect of agriculture on climate change, because the small-scale family farmers favoured by a local food system are more likely to practise low-impact farming. This is why sustainable agriculture goes best with a local food economy.

Compare that to how most food is traded and sold.

How food is usually traded today stands in the way of a sustainable food system. When trying to figure out how to feed nine billion people in the next few decades, within the environmental constraints we face, we must understand what kinds of markets best support a sustainable, resilient food system where all people can afford to eat healthy foods and where farmers are paid a living wage to grow food in a way that has the least consequences for the environment. In fact, the definition of sustainable agriculture that is commonly understood goes beyond growing methods to include the economics of food. It is in local and regional food systems that communities can thrive, in both the developing world and in wealthy countries, and where farmers can earn a living while the public has ready access to healthy food. To be sustainable, we need local food systems that are good for the environment and for local economies too.

We saw in 2008 how vulnerable our global food supply chain is to price shocks and what this meant for people everywhere. Around mid-year, it suddenly cost a lot more to eat. The wheat that is turned into chapatis and flatbreads and loaves in homes and bakeries around the world jumped in price. The cost of rice in markets and grocers and supermarkets went up dramatically too. This was because on world commodity markets, wheat was trading at 130 percent more than it had only a year earlier, while the price of rice was 70 percent higher. According to the FAO's Food Price Index, which measures the average cost of food by taking stock of commodity prices, the cost of wheat, rice, and corn peaked well above what it had been only two years earlier. Tens of millions of people, mostly in developing countries, couldn't afford to eat as much as they had just a few months before. The situation came to be known as the 2008 food crisis, and in December of that year, the FAO announced that the higher food prices meant there were forty million more undernourished people in the world than there had been only a year earlier.

There has been a lot of analysis and explanation of what happened in 2008, as well as of what has transpired since. Many in the media initially blamed the growing livestock industry that was eating up, literally, grain stocks, as well as a soaring global appetite for meat and biofuels made from corn, oil palm, and sugar cane. Yet several years later, a consensus has emerged linking 2008's surge in food prices directly to the global financial system. In the words of Olivier De Schutter, the United Nations special rapporteur on the right to food, a new global food trade regime exists today that is characterized by "the intertwining of food, energy and finance, changing global supply and demand dynamics and greater consolidation in the agri-food sector."[16] Food has rapidly become linked to entities in the global economy where there is no concern for how food is produced or who has access to enough to eat.

It wasn't like this in the 1980s and 1990s. This international system that trades in food is brand new and is having a direct influence on agriculture and on people's ability to feed themselves—what the United Nations calls their human right to food. The way things work today runs counter to our goals. How governments and international institutions regulate food today supports industrial agriculture. It also creates a particular type of marketplace in which certain farming practices and economic models are given the opportunity to thrive while raising hurdles for more sustainable alternatives. In order to figure out how to make a food system that is fair for farmers and that feeds us all, we must learn from the problems inherent in today's system.

The industrial food system has made it even more difficult for the family farmers of the world to earn a living. Farmers everywhere are the poorest segment of the population. In the developing world, these are the people who eke out a living from the earth by growing what they can, often going hungry because they don't produce enough or earn enough to nourish themselves. That is not to say that the feudal arrangements of history that saw vast numbers of peasants toil in their fields for the benefit of their overlords was any better. But rather than creating fair food economies in which the people who produce the food have the chance to make money, the industrial food system has empowered new overlords, and the farmers continue to toil, saddled with debt and little or no profit.

In the developed world, the majority of farmers aren't doing well. The price farmers are paid for what they grow hasn't risen, however at the supermarket, the cost of food has gone up, meaning farmers are getting less and less out of every dollar the consumer spends on food. Around the time of the Second World War, a farmer earned fifty cents of every dollar spent. In 2006, according to the US Department of Agriculture, that had dropped to 19 cents. So farmers have struggled to become more efficient and increase their economies of scale to make a profit, but all they have ended up with is a growing gap between their costs and their income. In the United Kingdom, it is estimated that the move to big farms has meant a 61 percent decline in income from farming over the last thirty years. Certain kinds of farmers—such as livestock and dairy—have struggled more than others. And "usually small and mid-sized farms fare the worst," said Alicia Harvie, program manager at Farm Aid, a Massachusetts-based organization that has run a crisis phone line for American farmers since the 1980s. "We get calls from farmers in crisis every day." Some big farms growing commodities, such as corn in Iowa, in recent years have benefitted from rising crop prices. However, in 2012, when what the USDA called the most severe drought to have affected American agriculture in 25 years destroyed corn crops in the Midwest, it was crop insurance, rather than sales, that generated the income on large farms.[19]

How we got in this mess isn't straightforward.

Ever since the early twentieth century, governments have treated agriculture differently than other sectors of the economy. It has been understood that growing food is not the same as, say, making widgets. One difference between food and other commodities is that the demand for food doesn't tend to change much. Food has what economists call a relatively inelastic demand curve: everybody needs to eat, and they will buy food regardless of price.

Secondly, a lot of risk is inherent in farming that doesn't exist for businesses that make clothing or bicycles or computers. Farmers thrive and fail at the whim of the weather. If Mother Nature brings too much rain one year—or too little of it, triggering a drought—the harvest is compromised and the farmer loses income. But even when the weather is perfect, the farmer doesn't necessarily profit. If growing conditions are ideal and farmers reap bumper crops, then the flooded market pushes down the commodity price. So unless a farmer increases his productivity while his neighbours don't, he won't make any more money. This is called the Paradox of Plenty.

It wouldn't be good for society if farmers went out of business, so over the last hundred years, governments around the world have typically stepped in to try to help with this market failure. In the United States, the federal government introduced modern agricultural subsidies in the 1930s, with the New Deal and the Agricultural Adjustment Act. Subsidy programs included a range of payment schemes including income supports, whereby the government paid farmers an inflated price for their products. Today, it is the Farm Bill that governs agriculture and food programs and some farmers still receive lots of government help—as was the case for corn farmers who saw their crops fail in the drought of 2012. Agricultural subsidies are also commonplace in Japan, Korea, and the European Union. In fact, developed countries allocate so much of their resources to supporting domestic agriculture that it has been calculated that government subsidies to farmers amount to $1 billion *a day.*[20]

But this money tends to reward the biggest farms and the richest farmers, rather than providing support for small farms. According to investigative research into the American subsidy system by the not-for-profit Environmental Working Group, 75 percent of farm subsidies in the United States are handed to 10 percent of farmers—and these are the richest farmers.[21] It's a similar story in Canada. In 2010, only 3.2 percent of Canadian farms had sales of $1 million or more, but it's these farmers who received 20 percent of government subsidies. More than a third of subsidies support farms with gross revenues of over $500,000. The majority of Canadian farms, with substantially lower incomes, benefited the least—or not at all.

Government farm support programs tend to reward operations that are the least resilient, such as the highly specialized operations of industrial agriculture where the farmer focuses on one product. In these cases, all of the farm's eggs, so to speak, are in one basket. That's because these larger operations either produce for export or deal in commodities to sell into the agri-food economy and are highly susceptible to the wild swings of the world commodity markets. So when prices are down for what they specialize in and they don't earn what they do in a good year, they qualify for government subsidies to make up the difference in their incomes.

For the small organic farm down the road from the cash-crop operation or the industrial hog barn, there isn't crop insurance. Besides, since it is less likely that diversified small farms would experience catastrophic crop failure in the first place, because they actively reduce the risk by producing a variety of products and are typically not involved in the commodity markets, government supports are irrelevant to them. "The safety net leaves out anyone who is not towing the line," said Harvie.

Another problem with subsidies can be that when farmers are paid higher than market price for their grain by the government, they tend to produce more and more. This too favors industrial methods of production and also floods the market. The surplus has been sold internationally or even donated to poor countries in the form of food aid. These subsidized food grains from developed countries end up in the village market, usually at a lower price than what local farmers are selling their produce for. The local farmers can't compete and are often put out of business. Therefore, tax dollars spent on agricultural subsidies in rich countries end up pushing small farmers in developing countries off their land or further into poverty. The consequences of agricultural subsidies in developing countries have been so severe that everyone from the World Bank and the OECD (Organisation for Economic Cooperation and Development)to Oxfam is calling for reform.

It's more than government policies that shape agriculture today. The way the international monetary and trading system has been set up similarly supports industrial agriculture over sustainable agriculture, and large-scale commercial trade in food rather than local, small-scale food chains. International trade is governed today by the World Trade Organization, the global community's youngest multilateral organization, founded in 1995. (When the WTO was created, it set up shop in Geneva, in the old World Labour Organization building. Part of the renovations to prepare the space included covering over the labour-themed murals that depicted workers in their glory, an insider told me. Several years later, when the WTO had gained its stride, this art was resurrected, he said.) Almost every country in the world belongs to the WTO and each member has a vote, with some countries, such as Syria and Iran, holding observer status. North Korea and Somalia don't belong at all. Unlike at the United Nations, no one country has veto power. However, as a government negotiator explained to me, that doesn't mean that the rich, powerful countries of the world don't dominate.

The WTO governs all global trade, enforcing the rules everyone has agreed on to ensure that business runs smoothly. The goal is to liberalize trade by removing barriers such as tariffs and quotas that get in the way of trading manufactured goods like cars and textiles and cellphones so that economies can prosper. Domestic policy in member countries is not supposed to distort trade. Rather, policy should be market oriented. For example, import barriers that restrict the amount of goods that can be imported into a country are considered trade distorting. When a state joins the WTO, it agrees to ensure that its domestic policy respects the international laws of trade.

This system is based on the belief that trade works best if governments get out of the way of the market. The idea is that prices act as signals to consumers and influence people's decisions to buy one product over another. Whether or not a consumer purchases something sends a signal back to the seller, who can adjust the price accordingly. Economic theory says that when governments get in the way of these signals by creating trade-distorting policies, it interferes with the natural functioning of the market.

The World Trade Organization has worked to liberalize trade in all sectors to allow an unfettered market to operate. Up until the 1980s, food and agriculture had been left out of trade liberalization by governments brokering free-trade deals. But agriculture has come to be included under the WTO's jurisdiction, though it does remain the most protected sector and so far hasn't been subject to the same degree of liberalization. The reasoning behind liberalizing agricultural markets is that if a country opens agriculture and trade, theoretically that will lead to increased production. And when a country exports its food to the international market, it creates on-farm jobs, people's incomes rise, and increasing business opportunities encourage more land to be brought into farming.

Following this logic, countries all over the world, but particularly in

Africa and South America, have created policies to trigger a shift away from producing food to feed the local population, and instead towards growing products that can be sold internationally. In the 1980s and '90s, the World Bank and the International Monetary Fund required many developing countries to make macroeconomic changes to their economies to precipitate this transformation as part of so-called structural adjustment programs. These were market-liberalization policies that were required in exchange for new loans or a better interest rate on existing loans. Structural adjustment was a complex package of ideas that affected various sectors, including agriculture. It forced the privatization of Crown corporations, the deregulation of markets, and the elimination of supply management systems. There has been furious debate about the effects of structural adjustment, and an immense amount of criticism because much research has found that the policies worsened the poverty they were intended to help. Structural adjustment is largely responsible for a shift in government policy in many developing countries away from supporting a system of small farmers producing for domestic consumption. Instead, policy was made to support export agriculture that was supposed to bring in foreign currency. Such policy shifts have created obstacles to sustainable food systems capable of feeding people.

According to a 2011 report by the UN's Olivier De Schutter that examined the influence of the World Trade Organization on food security in developing countries, when domestic agriculture is overlooked in favour of export-oriented production, states lose the ability to provide food for themselves. This loss was palatable to governments from the 1970s to the mid-2000s because international commodity prices were historically low. On paper, it made economic sense for a developing country to buy cheap food grown and subsidized elsewhere and then specialize in export commodities such as coffee, fresh fruits and vegetables, and cut flowers to sell for foreign currency. An outcome of these policies was a restructuring of international trade and food markets. Between 1990 and 2004, agricultural trade doubled and is now valued at over $1.1 trillion (processed foods constitute a whopping 80 percent of this trade).[22] But when food prices shot up in 2008, governments were sharply reminded of the consequences of relying on someone else for food.

Agricultural trade might be booming, but only the handful of corporations that trade in most of the world's food are benefiting. Developing countries aren't reaping the winnings, and this means smallholder farmers on the whole aren't profiting from the worldwide food economy. In fact, developing countries have gone from being net food exporters to net food importers, and the least developed countries—the poorest countries in the world—now import, on average, 20 percent of the food they need. Because commodity prices have gone up so much, the least developed countries watched their food bill rise by a third in only one year. According to De Schutter, when countries with good agricultural capacity depend on international trade rather than domestic food production, it can seriously limit the accessibility of food.

But as Chandrakalabai's story demonstrates, local food systems in which small and medium-sized farmers produce food in a sustainable way can be the catalyst for a new economy in which environmental protection is valued and farmers can start to make a living.

CHAPTER FOUR

Local versus Industrial: The Alternative Economy of Food

It takes about twenty minutes for Chandrakalabai to walk from her house in the village to her fields on the outskirts of town. But she doesn't mind the walk along the rutted dirt road, past the sugar cane fields. The day she took me there, the sky was blue, the sun was strong, the air was still, and there was the smell of smoke from fires built by farmers who were burning invasive weeds. Butterflies fluttered and birds flew over her crops of garlic, pigeon peas, cowpeas, and green peas, as well as alfalfa, or lucerne grass as it is known locally, grown for animal fodder. For the winter, she would soon plant chilies, spinach, fenugreek, green beans, and coriander. In one section of land, she and her husband had planted sweet lime trees that, in a few years, would produce a table fruit that fetches a good profit in the city. She also grows cotton and sugar cane, two irrigation-thirsty crops that wouldn't take well to the arid land here without the water from the bore well she paid to dig. "When there was no water, there was nothing here. Now it feels good to see things growing," she said. Around her fields, she and her husband built earthen banks to keep this irrigation water from running off, a blue plastic pipe distributing the water. We walked the rows together.

“This plant is called tulsi,” she said, pointing to a bushy plant with a purple-green leaf growing at the side of the field; it is also known as holy basil. “It helps in pest control.” She picked a bean and handed it to me to try. It tasted unlike any green bean I had ever eaten raw, its flesh firm between my teeth, the flavour almost bitter. Then on to the vermicompost pits, where her husband pulled out a handful of wriggling worms to show me the creatures that produce the fertilizer to help their crops grow. Standing a few yards away were three cows, tied at the neck, grazing on the grasses at the side of the fields. They appeared docile, and Chandrakalabai called them by name and explained that they belonged to a local breed of cattle that thrived in the harsh conditions of dry, hot Maharashtra. And all around their plot were piles of burnt Congress grass, an invasive weed named after the national governing party in the 1970s. The Congress government had accepted food aid from the West that arrived, unbeknownst to them, with the seeds for the noxious weed that now grows everywhere. The organic farmers must yank it out of the ground by hand and then burn it.

Chandrakalabai told me over and over how organic farming has changed her life.

With the income from this property, she has sent her son to school (he is now a teacher), purchased land to build her kitchen house and acquire more acreage, and paid to dig the well that irrigates her crops. When I asked her how she felt about her accomplishments, she looked at me and said: “I feel there has been a sunrise in my life.”

When Chandrakalabai told me about her sunrise, we were sitting under a tree hiding from the bright sun, just a few metres away from the headquarters of the Alexander Mahagreen

Producer Company. The company was founded in 2011 by a group of small farmers, among them Chandrakalabai, who had learned about organic farming from the IIRD and created organic farmers' clubs, which helped to spread agricultural knowledge and farming skills. Out of the clubs grew a number of microenterprises. Some women produced organic pesticides made from the neem tree they sold to farmers in their communities, while others made a biodynamic organic compost from manure, called Cow Pit Pat, that they sold for a profit. Still others invested in machines to grind flour, coconut, or chilies for masalas or to make the vermicelli noodles used in some traditional dishes. But as small farmers, scattered in villages across a large rural area, they didn't have access to a sizable market. So a group of ten women came together and decided that if they could collaborate to market their produce, they could potentially increase sales. They formed a producer company, a legal entity in India that is similar to a co-operative and can be made up only of primary producers such as farmers or fishers. "The idea comes from the farmers," said Navnath Dhakane, a program officer with the IIRD who acts as the CEO of Mahagreen. "The number of organic farmers had increased so much. The organic bazaar is only once a week and only twenty-five farmers can go there at a time. But now we have more than four thousand women in the area who are organic farmers. They were thinking where else can we sell? There is no other market for organic produce around here."

They took inspiration from the Amul dairy farmer co-operative, in the state of Gujarat, which has been ranked one of the top twenty dairy processors in the world. The Amul co-operative was founded in 1946 and today is made up of seventy thousand village co-operatives across India that bring together milk from thousands and thousands of mostly small—some landless—rural farmers. The co-operative today is worth $2.5 billion. It has succeeded by connecting the producers with one another and providing training and investments in the villages, something Mahagreen can emulate.

More than a thousand small farmers own shares in the Mahagreen Producer Company, and the hope is that it will grow beyond four thousand farmer shareholders. The company aggregates produce from the various farms and then sells it to large buyers in big-city markets. For instance, it sells pulses such as chickpeas, kidney beans, and lentils; grains like wheat, millet, and sorghum; and spices, onions, ginger, garlic, and more. Its customers include a supermarket in Hyderabad, a city about five hundred kilometres away. It also has buyers in Pune, Mumbai, and Bangalore. "Pune even comes and picks up in a truck!" said Dhakane with pride. "This week we sold 12,100 rupees"—that's more than $200 worth of produce in 2012 figures, a significant sum. Yearly they are earning about $9000. The hope is that access to these large markets will help to improve the standard of living in the rural areas and help to show that organic farmers can earn a decent wage from their work. "We want to create this as a brand," explained Dhakane.

Mahagreen's shareholders are the women farmers; a smaller group sits on the board of directors. Dhakane provides office help. "When we went to the bank to open an account, they asked us why," said Hapizabee Putham, one of the women farmers who acts as treasurer. "We said, we're starting a company, and they just laughed at us."

"Now three to four people want to come and join us every day," added another board member.

"Now no one laughs."

The women who run Mahagreen believe there are benefits for both the farmer and the consumer. "The farmer doesn't have to go to the market herself. They save that cost," said Radahbai Kalyan Shelke, the director of the company. "There's no middleman now. Even the consumer gets fresh produce. The food doesn't go from here to there to there and there. It goes directly to the consumer."

In Maharashtra, the benefits that derive from a local food economy are tangible. The IIRD has found that in villages where the majority of farmers use organic methods, there is a rise in small farm-based enterprises and prosperity. For example, there are an increasing number of businesses such as micro dairy and poultry operations, as well as of enterprises that add value to a raw product. Chandrakalabai, for one, bought a machine to grind coconut that she could then sell. Some villages have seed banks where farmers save their seeds to plant the following year and so avoid incurring the costs of buying new ones, among other benefits. Residents have also organized community elder care programs, including grain banks to which farmers donate extra food to feed those who are too old to provide for themselves. And in a country where more than half the population doesn't have a toilet—people in rural areas typically rise early to defecate on the outskirts of town before the sun comes up—organic villages have a higher number of toilets. In Chandrakalabai's village of Dhangaon, where there are ninety organic farms, out of 272 households, 175 have toilets. One of the women who is a member of the producer company told me in a group meeting that when they look for a husband for their daughters, they only look for an organic farmer. "Those houses are the ones with the toilets." All the women in the room laughed.

The producer company and the organic bazaar are just two examples of new market arrangements for selling food that people are creating around the world to support their local food systems. These market arrangements are different from the way most food is sold today because farmers have rethought the food chain. They have identified where they have a problem and tried to fix it by retaining control of both their inputs and the marketing of their produce. This control is most possible in a local food economy. When the food chain is shorter, the farmers are more likely to benefit because it is simply easier to sell directly to the customer if she is your neighbour. When the food chain is long or global, as we see today, farmers have a greater chance of foundering because they have little choice other than to rely on middlemen.

However, short food chains have their drawbacks too. In India, middlemen who supply the regional markets often buy from the village farmers at a depressed price. So do the small-time middlemen who can be farmers themselves trying to earn more money on the side selling at village bazaars such as the one in Bidkin. This is why farmers must be empowered to retain control of the marketing of their produce as well. That's where co-operatives and alternative ways of selling food play an important role.

These short-chain alternative economic arrangements aren't suited to large industrial farms because big farms need big buyers (supermarkets, exporters, processors) to purchase their huge quantities of food. And the big buyers aren't suited to smaller-scale farmers either. The prices these companies offer are based on the large volumes produced by industrial farms. They pay a certain price for a high volume, but when that sum is broken down to a unit price, it isn't viable for a small farmer, who can't possibly achieve the same economies of scale.

Because the smaller family-run farms that do best in the local food economy frequently practise sustainable agriculture, "local" is often used interchangeably with "sustainable." This is something the naysayers of local food systems have criticized loudly. They often argue that food transported long distances can have less of an environmental impact than locally grown food and point to studies that have calculated, for example, that growing apples in New Zealand and shipping them to Britain has less of an environmental cost than storing British-grown apples for the off-season. They argue that this is proof of a fatal flaw in the reasoning that local food systems are better for the environment. They go on to conclude that we should let economics and comparative advantage, not concern for the environment, decide where it is best to grow food. However, even this economic reasoning is based on the logic of the same flawed market.

"Food miles are relatively easy to calculate. There is something about them that is intuitively appealing to people," said David Cleveland, a professor of environmental studies at the University of California, Santa Barbara, who has spent a long career studying sustainable agriculture. "It is a great concept but it has its limits." In 2011, Cleveland released a study examining what would happen if all fruit and vegetable consumption in Santa Barbara County were localized. He found that if all sales were to shift to the local market, there would only be a reduction in greenhouse gas emissions of less than 1 percent. The results were explosive. Angry local-food advocates accused Cleveland of working for Big Food. Then Big Food called. A multinational food company asked Cleveland if it could fly him to its headquarters to film a video in which he would explain why food miles were bogus.

However, everyone missed the point—as is typical of the food miles debate. For one, Cleveland candidly explained that his study only compared industrial food transported great distances to industrial food produced locally. To gain a true understanding of the environmental cost, we would have to examine "the total amount of impact of processing the vegetables [in the industrial system]," he said, and we must compare them to fruits and vegetables grown sustainably here.

Imported fruits and vegetables are usually processed more than those we buy in a local, sustainable food system. "What's the environmental impact of the canning process?" he asked. What about the refrigeration required to ship food over great distances? What about the packaging and the waste generated along the chain? Without this information, the analysis is incomplete.

"If you had a local food system where food was produced in a way that sequesters soil carbon, naturally maintains soil fertility, doesn't depend on toxic chemicals with optimized transportation and distribution, and consumers stored and prepared food in a way that conserved energy, you would have huge savings," Cleveland said. The reason this kind of system doesn't exist today, and the reason growing apples in New Zealand and shipping them around the world can appear to be environmentally neutral, comes back to markets. To economics.

The industrial food system is mature. A lot of money has been invested in technology to increase its scale of operations and improve efficiencies. There hasn't been any injection of funds and technology into the new local food systems that are springing up, so there isn't a level playing field. These differences, however, are often used to discredit sustainable food systems. For example, critics of local food often point out that when a small organic farmer delivers her few small boxes of strawberries in an old gas-guzzling pickup truck, her vehicle emits more greenhouse gases per unit of food than the large trucks that ship strawberries across the continent, in vast quantities, to supermarkets. There is an energy savings by aggregating enormous quantities of food and moving it in these big trucks versus the small farmer driving it around herself.

But let's dig down a bit. The organic farmer is driving an old pickup truck not because she prefers to burn lots of gas but because that's all she can afford when she grows organic strawberries along with vegetables and possibly some eggs and meat on her few hectares of land. Compare her vehicle to a truck custom-built to carry strawberries on long-haul journeys. This vehicle is high-tech, kitted out with extra suspension to minimize the bumps and thus bruising of the fruit, as well as refrigeration to prevent spoilage. It's worth at least $200,000. If the local, organic farmer, working her small plot of land, had that kind of money to invest in transportation, she'd probably buy a hybrid—and then pay off the mortgage on her land to boot. (The huge strawberry business can invest in this expensive technology because of the tremendous market that exists for out-of-season strawberries in climates where we have the local fruit only a few weeks of the year. There's a lot of money to be made satisfying people's craving for fresh berries with breakfast in January.)

These big enterprises also have other technological advantages over that small farmer. To support this long-distance berry industry, agribusiness corporations fund plant-breeding research so scientists can create generation after generation of new and improved varieties. They organize complicated planting systems that involve starting seedlings in a sunny, warm valley and then trucking them to higher altitudes, where they bear more fruit. They use chemical pesticides, including the highly toxic ozone-depleting methyl bromide to sterilize the soil before planting. Then, low-wage migrant labourers typically harvest the fruit. When local farmers in places where this strawberry infrastructure doesn't exist try to sell their berries in June, they have a hard time competing and have to fight for shelf space in the supermarkets—assuming the supermarket can even accommodate small local farmers.

Advocates of long-distance food would argue that this is an example of a well-functioning marketplace in which farmers in one region reap the benefits of their comparative advantage. Their geography and climate allow them to produce more strawberries for a cheaper price than farmers with shorter seasons, smaller farms, and probably higher labour costs. But while on the surface it might look like a smooth-functioning market, in reality the large strawberry producers of California are able to produce more fruit at a lower price because they are profiting from what James Boyce, an economist at the University of Massachusetts Amherst, calls an implicit subsidy.

Unlike many agricultural subsidies, "these are not direct government handouts," he said to me. "The implicit subsidies come from a failure to price things at their true cost. The true cost of fossil fuels, pesticides, fertilizers. The true cost of these inputs is radically understated in this market because the environment is not priced in." This explains why I can buy a quart of strawberries from California for $1.99 ($1 on sale!) that have traveled 3,025 miles (4,868km) to my city while local organic berries in season go for $6.00 a quart. The California berries may have been grown in soil fumigated by large machines spraying toxic chemicals; they may have relied on aquifer-depleting irrigation systems; they may have sucked energy while being refrigerated for weeks; they may have been picked by low-wage migrant labourers who work in poor conditions; they may have been transported thousands of kilometres across the continent. However, these berries cost a fraction of the price of the organic berries sold at my local farmers' market *because* of these implicit subsidies. This is a failure of the market as we know it to appropriately allocate costs to different products. It would be fair to assume that the California strawberries would cost a lot more than the local organic ones if the real costs to society and the environment were factored in.

It is precisely this malfunctioning market that is inspiring Boyce and other economists to argue for a new way of structuring our economic system so that we support sustainability rather than environmental destruction. Boyce is one of the founders of Econ4, an international network of economists who are creating an economics for the twenty-first century that departs from the orthodoxy of today's mainstream thought. "It seems to me farming and agriculture are a perfect example of how and why we need these principles to be put into effect," he said. "For a long time, economics has itself been a scene of intellectual monoculture. Just like a corn monoculture in the midwestern United States. We have a few varieties of a single ideology that is reproduced across the United States and increasingly elsewhere in the world."

The Econ4 network advocates for true-cost pricing that takes into account non-market values and includes a full accounting of the costs and benefits of producing something. They suggest that in an improved economic system, the goal wouldn't be to maximize efficiency and profits at one point in time as we do today, but rather to minimize economic vulnerability over a long time frame so that we build resiliency into the system. A third pillar of their economic philosophy is a level playing field they characterize as a fair distribution of wealth, which includes nature's services and bounty, so that everyone has an equal opportunity to participate in the economy. Finally, they posit that an effective economy that serves all of us as well as the earth needs to exist within the framework of a transparent, accountable democracy.

"If we are going to have an economy that works for people, the planet, and the future," Boyce told me, "we are going to have to have an economy in which the distribution of wealth is fair and a polity in which the distribution of power is fair. We need democracy. When one has more democratic distribution of power in society, it is not only good for people but it is also good for the environment."

The new economic models that the sustainable food movements are creating, such as the one in Maharashtra, are real-life experiments in how a new theory works.

How do we know that local food economies are in fact accomplishing what they set out to do? Returning to Chandrakalabai's story, we might ask what was at the root of the radical transformation in her life—and in the lives of the other small farmers in the villages who now have toilets and better marriage prospects for their sons. Was it really the money earned through direct sales of organic agriculture that improved their standard of living? Or perhaps that came about because women were introduced to microfinance and learned to save and invest in one another, providing access to much-needed credit? Or maybe the improvements can be linked to the farmers' proximity to the city and an urban market where people are willing to pay more for organics. Maybe they just worked really hard? Then again, perhaps there are more toilets in these villages because these particular farmers simply applied for government grants that subsidize latrines while other villages haven't enrolled in the program. Are organic farming and a local food economy at the root of change, or is it something else?

We know that in Chandrakalabai's case, her financial situation first improved when she stopped having to buy seeds, pesticides, and fertilizers from a local dealer and instead started to save her seeds and make her own inputs. Chandrakalabai also told me that she reduced her medical expenses after going organic because she no longer needed to pay for visits to the doctor as frequently for her skin conditions. Before, when she was handling agricultural chemicals, her health had suffered. "You put the fertilizers on the soil and then touch the soil, and your skin becomes itchy," she told me, her eyes growing bigger as she described it. "It was a constant itching. I would be itchy while I was cooking. Now we don't get that itch anymore."

The health of her soil improved as well. Whereas the land had become hard and dry when she used chemical fertilizers, after two years of growing without them, she noticed an improvement. "The plant coming out of the soil is the indication of soil health," she explained. "Take toor dal. The bean pods don't come out hollow and infected. They come out feeling robust and all the leaves are green." Since switching growing methods, as well as investing in an irrigation system, she has watched her yields go up. These higher yields mean more product to sell. And at the organic bazaar, which is only an hour away from her farm, she has access to a clientele who places value on her pesticide-free food and pays her more than what she could get from a middleman. As a conventional farmer, she explained, she had earned one lakh—100,000 rupees, or about $1800—spending 50,000 rupees on the inputs she needed to farm that way. Now her organic farm earns her three to four times what she used to make in a year, and she no longer pays the cost of the conventional inputs. "My farm has developed well!" she said.

Clearly, a multitude of factors explain Chandrakalabai's sunrise: higher yields, money saved by not purchasing seeds and fertilizer, the investment in irrigation, her proximity to an urban market where organic food earns a premium. What all these factors do together, though, is allow her, the farmer, to be in control of her business. She takes back some of the power she lost when she participated in the industrial food system. For example, when she saves her seeds and makes her own fertilizer, she is not beholden to any company's asking price. When she sells directly to the customer at the organic bazaar rather than to a middleman, not only does she keep more money but she decides on a price. And whereas a small farmer like Chandrakalabai wouldn't have access to affordable credit from a bank, with a village microcredit program she has the control and the choice to invest in her business so that it grows and has a chance to prosper. These many factors that work together to improve life are a benefit of a sustainable local food system.

This goes back to the point James Boyce of the Econ4 network makes about the importance of distribution of power in a society. For an economic system to work for people as well as for the planet, there needs to be a fair distribution not only of wealth but also of power. Sustainable local food systems transform the way we buy and sell food because they change the dynamics of the market—they change which economic players have power in the marketplace. This has a profound effect on society. So this model's propensity to benefit the people who typically lose out in today's economy transforms food and agriculture into poverty-reduction and rural development programs.

According to a world development report put out by the World Bank in 2008, GDP growth that comes out of agriculture is two times more effective at reducing poverty than GDP growth from other economic sectors. In his 2010 report to the UN Human Rights Council, Olivier De Schutter wrote: "Only by supporting small producers can we help break the vicious cycle that leads from rural poverty to the expansion of urban slums, in which poverty breeds more poverty." He argues that investment in agriculture is most effective when growth in the sector comes from a rise in the incomes of farmers like Chandrakalabai. And this in turn leads to more demand for the goods and services provided within local communities.

Because local food economies are beneficial on so many levels, governments in rural regions everywhere are looking to build them to bring life back to their communities. In New England, the importance of agriculture to the local economy is particularly clear. Vermont was once known as the breadbasket of the region because farmers produced so much wheat in the Champlain Valley. But these wheat farms, as well as most other farms, didn't last until this century. Over the past fifty years, the state lost 90 percent of its farms. This meant that people in Vermont relied on food from farther and farther away. However, things started to change in the past decade, and there has been a slow but steady increase in small farms and food artisans such as cheese producers—people building small industry around food. In 2009, Vermont approved the Farm to Plate Investment Program to increase economic development in the state's agriculture and food sectors, create jobs, and formalize a process that had already started at the grassroots level. The state created a ten-year strategic plan to reinforce the policies and investments needed to build the strong local food systems it envisioned. Simply put, the goal is that food will nurture a vibrant state with a healthy economy. Investing in food and agriculture strengthens the economy because of what's called the multiplier effect—that is, an increase in spending that causes an increase in income and consumption that is greater than the initial amount spent. According to a report by the Fair Food Network about the economic impact of localization in Detroit, if 20 percent of people's food budget there was shifted to local business, it would result in more than half a billion dollars in local economic activity, including 4700 new jobs. The city would also collect $20 million in business taxes.

Let's track how the dollars circulate—and maybe even multiply—in the local food economy.

When the consumer buys, say, a dozen eggs from a farmer at the market, the farmer now has money to spend on chicken feed grown by the guy down the road, or on bread, or on whatever else she needs to buy. As the money the farmer earned selling eggs at the farmers' market circulates, other people in the community start to make more products. Maybe the farmer down the road now can produce more grain to sell for chicken feed. People can even start to make *new* products. Let's say someone opens a bakery where the farmer can go to buy bread, or someone starts to make cheese. Maybe a young chef moves back to the region to open a restaurant. And as this local food economy develops—the cheese, the bread, the chicken feed, even the culinary tourism opportunity that the chef's new restaurant presents—it attracts demand from outside of the community too, which in turn increases the amount of money moving around the community.

In Maine, just as in Vermont, there has been a good food explosion. New restaurants have opened, chefs are creating a cuisine that reflects the geography of the state, new farmers have arrived to practise sustainable agriculture, and people have started artisanal food businesses. A study published in the *Maine Policy Review* looked at the economic contributions of the state's food industry—farmers, food processors, food sellers, and so on—and tracked how much economic activity was stimulated by each dollar in sales. What it found was that every $1 that Maine's food makers earned in revenue generated an estimated $1.82 in statewide economic activity. And for every one person working in the sector, an additional 1.2 jobs were supported by these food-making activities. The conclusion was that the food industry contributed $11.5 billion in sales revenue to Maine's economy. "If you look at what happens to the money, fewer of the dollars go out of the community and instead stay cycling in the community," said Linda Silka, director of the Margaret Chase Smith Policy Center, publisher of the *Policy Review*. "And that's one reason why small farmers are so important. These small farms also act as anchor institutions. They are important to their community"

The reason the multiplier effect works particularly well for food rather than, say, the manufacturing of cellphones is because everyone has to eat, usually at least three times a day. "A lot of these smaller communities, they were never manufacturing hubs, but people lived there and they got their food there," said Armine Yalnizyan, senior economist at the Canadian Centre for Policy Alternatives. "There is this underlying infrastructure." An infrastructure that is, in many cases, lying dormant, ready to be revived.

Local food economies can help cities too, particularly when it comes to urban agriculture. Cities have long depended on food produced within their borders. While we in the West have become disconnected from this tradition, elsewhere an awareness of the importance of urban agriculture to urban food systems and economies is growing. In most cities in Africa, there are hives of activity centred around food production. In Ouagadougou, the capital of Burkina Faso, 44 percent of the city's population grows vegetables or raises fish or animals to sell. In Dar es Salaam, Tanzania, urban agriculture is the second-largest employer. In Kampala, Uganda, city food production is significant enough to be regulated by the city government. People grow vegetables and fruit in the city as well as other foods. Small urban dairies supply big East African cities such as Nairobi and Arusha with fresh milk. In West Africa, people keep pigs and poultry in places such as Yaoundé, Cameroon. Urban food production is similarly important in Latin America and across Asia too. In Brazilian cities such as Belo Horizonte and Brasília, vegetables as well as livestock are produced within the urban borders. According to the RUAF Foundation, an international urban agriculture institute, 80 percent of the fresh vegetables in Hanoi, Vietnam, are produced in the city, along with 50 percent of the pork, poultry, and fish and 40 percent of the eggs.[23]

Diana Lee-Smith, a social scientist affiliated with Ryerson University's Centre for Studies in Food Security, has been studying urban agriculture for decades. She lives in Nairobi, Kenya, a city whose poorest residents depend on urban agriculture for their survival. Lee-Smith is one of the founders of the Mazingira Institute, an independent research and development organization that focuses in part on urban agriculture in Africa and that carried out the first survey of city farming in Kenya, in 1985. She has long seen the importance of urban agriculture in the city that many others couldn't. "It was so blatantly obvious. If you moved around any city, you could see urban agriculture, if your eyes were open. But most people don't because it doesn't fit into their categories of thought," she told me. When Lee-Smith and her colleagues completed their survey of what was being produced in Nairobi in the 1980s, they were able to quantify the sector—much to the amazement of others. "People were baffled," said Lee-Smith. "They thought agriculture will die out because it's a relic of rural behaviour and will disappear."

Quite the opposite. People grow food in urban areas not out of habit but because they are hungry and they don't have money to buy enough to eat. They also produce food in the city because of a lack of infrastructure. For example, in developing countries in the south, there isn't access to refrigeration, making it impractical to transport perishable foods over long distances. When perishables such as salad greens and milk are produced within their markets, they can be sold fresh. Also, there is a persistent crisis of unemployment in the megacities of the developing world, and urban agriculture is a way to make money. Selling food to your neighbours is good business. Lee-Smith said, "Most people—and we confirmed this with the survey—they are going to feed themselves and to hedge themselves against starvation." But when they did a strategic analysis, they found that, as a proportion of the overall population, more high-income than low-income people are doing urban agriculture. "They are the ones making a living off of this," Lee-Smith said. "What I don't know, nor does anybody else, is whether the poor get rich by doing urban farming. I certainly think so." The microindustries of urban agriculture play an important economic role.

In Nairobi, farmers in the city produce just about everything. Fruit and vegetable growers sell papayas, amaranth, bananas, eggplant, cilantro, tomatoes, lettuce, kale, spinach, and more. Women farmers in the neighbourhood of Kibera, the city's largest slum, have created what they call vertical farms. These are large sacks filled with soil in which the women plant their vegetables, slicing holes into the side of the bags to maximize the vertical growing space. Other farmers grow more traditional crops on idle public lands with tacit agreement from the authorities.

Of course, urban agriculture in the city is not restricted to the slums, and people living in middle-class neighbourhoods everywhere are also involved. They keep cows and goats for milk, chickens for eggs and meat, as well as hogs. Some women have started value-added enterprises and make peanut butter, jams, and pickles to sell. "This is increasing," said Lee-Smith. "The proportion of people doing agriculture is increasing faster than the rate of urbanization. What does this mean? Part of it, I think, is you can feed yourself and you can even make a living. If you are in Dar es Salaam and you have a cow and a few hundred square metres, you are well off. That's because they are sitting on a huge market."

In Nairobi, a growing number of people are choosing to invest in their own urban agriculture businesses. In the Riruta neighbourhood where Lee Ngugi Kamau grew up on the outskirts of the city, he has built a small rabbit meat enterprise he calls Weru's White Meat. In a well-kept outbuilding, he keeps more than a hundred rabbits that he buys from a women's co-operative in the countryside, about twenty-five kilometres away. Kamau fattens the animals before he processes them in his own slaughterhouse and then sells the meat to supermarkets and restaurants in the city. Before starting this business in 2009, he kept the rabbits on a more informal basis and worked in the city's hotel industry. He went into urban agriculture because he wanted the opportunity to run his own operation. "We've actually seen a shift in Nairobi where young people in different careers are getting into agriculture," he told me from Nairobi, where there is a high youth unemployment rate.

At thirty-two, Kamau is part of the trend. Some food entrepreneurs collect milk from the many small livestock farmers in the city to turn into yogurt that they sell on the informal market. Others are growing sprouts that they supply to supermarkets, and some are even raising guinea pigs for meat, a food that hasn't typically been eaten in Kenya but that comes from an animal suited to urban husbandry. One of Kamau's friends grows green vegetables that he both sells to grocers and delivers as part of a food basket to clients in wealthy parts of the city. Another raises fish in ponds on his land on the city outskirts and supplies the fish to restaurants. A third friend grows passion fruit and sells juice from his harvest. These are just some of the many local food businesses in the city. "Most youth with their innovative and creative minds see that agriculture can make money and increase food security," said Kamau. "From my perspective, I think that urban agriculture is a revolutionary thing because the impression here is that to be a farmer you have to have a large tract of land. People are realizing that you can grow your own food in your backyard."

As more and more of us move to the city, this kind of activity will grow in importance. By 2025, more than half of the African population will be urban, and the number will continue to rise quickly. Yet urban farmers in developing countries face many hurdles. Often, people don't have security of land tenure or the access to credit that would allow them to invest in their businesses. Vegetable producers in particular face health risks when they tap into wastewater sources, such as streams and effluent or even sewers themselves, to irrigate and fertilize their crops with all that nitrogen in the human waste. However, local governments have a role to play in improving the context for urban agriculture. Wastewater, for example, could be treated first. Then the urban farmers would have a good source of fertilizer and they would also be providing ecological goods and services for the city by recovering nutrients that would otherwise have been lost.

Urban agriculture invites us to think creatively about how we can improve our local food systems and strengthen our urban economies too. And just like their rural cousins, urban farmers can take control of their destiny.

When people start to have control over their food and over their lives, the feeling of empowerment is contagious. A shift in who holds the power and wealth in a society has wide-reaching ramifications. According to Jules Pretty of the University of Essex, when farmers thrive in vibrant sustainable local food systems, people are inspired by what they accomplish and this confidence spills into other parts of their lives.

Women in particular stand to benefit. At the meeting of women agricultural extension workers where I first met Chandrakalabai, a smiling woman sitting at the front of the group wanted to tell me her story. She was from a village that followed a particularly conservative interpretation of Hinduism; women were not permitted outside, even to work in the fields. "Women don't step out of the house," she told me. "We wouldn't ever speak to men, and we always have to put the sari over our heads." Then, fifteen years ago, she became an organic farmer, growing foods such as sweet lime and onions that she sells at the organic bazaar as well as from her home, to neighbours; she also grows sugar cane and cotton, and earns extra money teaching other farmers. After she started her enterprise, things changed at home. "My husband was really short-tempered. Now if I have to go to a remote village"—to teach other farmers—"he takes me." In her village, the water supply is turned on only once or twice a week, for one hour, and village women must rush to collect as much as they can from the communal pipes. "Today as I was leaving, the water came on," she said the morning of our group meeting. "My husband said, 'Go ahead, you'll be late, I'll fill it.'" The idea of a man collecting water so that his wife could attend a meeting made the whole room laugh with glee.

Chandrakalabai had a similar tale of how taking charge of the food system led to other changes. A few years after she started working in agricultural extension, travelling around the area teaching other farmers about organic methods, her mother-in-law encouraged her to run for elected office. Each village in India has a sarpanch, a village head, and a council, the panchayat. This is part of the village democratic system as envisioned by Gandhi, created after independence from Britain. In the early 1990s, the government reserved a third of sarpanch positions for women, making it easier for females to be elected in a society where they typically didn't have leadership roles. When Chandrakalabai went campaigning, all the village women complained to her about water. Dhangaon didn't have municipal water service. Her village had no running water at all. Every morning before dawn, the women would walk to the outskirts of town with their aluminum water pots and line up to hand-pump enough well water to last the day. They wanted running water, they told her. So Chandrakalabai promised them running water.

The morning after she won the election, she gathered all the women together. "I said, 'You want water, then bring your empty vessels and meet on the road,'" she told me. "One woman from every house came. At first we just sat there on the road and blocked traffic. Then the member of the Legislative Assembly showed up and they said, 'We will arrest you.' We said, 'Sure, arrest us!' So we all went to the police station—two truckloads of women and one truck full of men. We sat under the trees at the police station. They took all of our names and noted what we were wearing. We got hungry and we asked them to feed us. So the politician asked for *pava*"—a rice snack—"for the women from a nearby restaurant. They served it to us on pieces of paper. Then we ate and they told us to be patient and they will make water available to our village. But the women said, 'We're not leaving.'" The politician finally promised they would have water service in Dhangaon within the month. And they did.

I'm not arguing that sustainable local food systems will solve all problems—bring water to villages, empower women, enrich the poor—though as we've seen, sometimes they can have this effect. There are many hurdles in India to ensuring that a local food system can feed people and allow farmers to earn a living wage. Most villages don't have regular electrical services to power irrigation pumps, *if* the water pipes have been connected to a local supply or money has been spent to dig wells. The roads are bad, which makes it challenging for small farmers to get their food to market. Because of poor food handling in the country, the government believes that about half of the food grown is spoiled before it reaches the market. Corruption in politics is rampant, which makes fixing these problems even more challenging. There is no panacea.

What I am arguing, however, is that sustainable local food systems offer us an opportunity. They provide a path to follow that has the potential to start improving the situation of small farms everywhere. By helping to raise the smallholder farmers of the world out of poverty by designing a new kind of marketplace, we will be on our way to creating a food system that is not only more sustainable but also more effective at feeding us all. This new economy of food is an important part of that first step of ecologically recalibrating agriculture and deindustrializing the food system by 2050.

The next step has to do with labour.

Chandrakalabai's story demonstrates how sustainable food systems function in the developing world where many people still live in rural areas, their livelihoods dependent on their small farms. But the same currents are at work thousands of kilometres away from Dhangaon, on the other side of the earth in North America, where we've made the transition away from an agrarian society. Here, too, the question of who will work on our farms is starting to be answered.

CHAPTER FIVE

The Twenty-First-Century Peasant: But Who Will Grow Our Food?

It was the third day of September on the sand plain in Norfolk County in Southern Ontario, but the late-summer sun was hot. It felt like July. Fall was nowhere in sight. Along the fencerows, the flowers were in bloom. There was the purple of echinacea blossoms, and the yellows of the lanceleaf coreopsis, the brown-eyed Susans and the ox-eyes. A herd of cattle ranged on the grass. The flies swarmed, and as the cows chewed their cud, the animals flicked their tails, paying us little notice. At the back of Bryan Gilvesy's farm, an old watering pond looked like an inviting place to stop and swim. But we still had a lot of land to cover. After feeding me a lunch of the farm's own beef, smoked on the barbecue, with tomatoes from the garden and bread, Bryan was taking me on a high-energy tour of the property, powered by the enthusiasm he and his wife, Cathy, have for the new approach they've developed for raising cattle in this ecosystem. It is one thing to make the case that sustainable agriculture can improve life for small farmers in India, where poverty, population density, a lack of education and access to technology, and a scarcity of land have prevented modernization and industrialization of farming on the scale we've seen in the West. I'd come to visit Bryan and Cathy to see first-hand what a thriving sustainable farm looks like in a country where lots of machines and not many people work hectare upon hectare of monocrops.

Until recently, tobacco was the mainstay in Norfolk County, on Bryan and Cathy's farm as well as their neighbours'. The farmers grew tobacco under a supply management quota system that kept prices stable and made for decent profit. A family could do well on a hundred acres, rotating the tobacco crop with rye they used as a green manure to maintain fertility and build organic matter in the soil. The earnings from a farm that size paid for the average family to send their kids to university as well as to save enough to retire. The area has attracted generation after generation of farmers, and tobacco had long been the crop of choice. American settlers arrived in Norfolk County in the 1930s, setting up plantation-style tobacco farms. European farmers arrived next, from Belgium, Germany, and Hungary—both Bryan's and Cathy's grandparents were Hungarian-born. Bryan's grandfather was a sharecropper before he was able to buy land of his own.

In 1979, when Bryan was nineteen and just out of high school, his father encouraged him to take on what was called a quarter-crop mortgage. Bryan bought a farm with a down payment he borrowed from his parents and agreed to pay off the rest of what he owed to the previous owner by giving him a quarter of his crop every year. This was how older farmers passed their farms to the next generation. So Bryan started to farm and, since the tobacco came off the fields by mid-September, he still was able to go to university to study business. After he finished his degree, he worked full-time on the farm, where he did well. He bought the neighbour's farm too, got married, and had kids. But something wasn't right. "Tobacco ran us into such a ditch that we re-examined everything about our lives," Bryan told me. Already in the 1980s, the tobacco market had begun to falter. The cigarette companies could buy the commodity crop at lower prices from Zimbabwe and Brazil, and anti-smoking campaigns were putting a question mark over the future of the industry. "First of all, we wanted to get off some major corporation running our lives," said Bryan. "Secondly, we wanted to get off the chemical companies making money, no matter what. Then we wanted to get healthy. It scared the crap out of us living in that environment with all the chemicals. That style of farming just seems insane after a while."

In 1993, the couple made a decision that would eventually change the course of their lives. Just for fun, Cathy and Bryan bought two mother cows—Texas Longhorns, with big curving horns. The animals look like they've stepped out of an old western movie. Until then, the couple hadn't had any livestock. "It turned out we had a very good eye for cattle," said Bryan. They renamed the farm Y U Ranch, and within ten years, half of the farm's income came from breeding the Longhorns for the American market, with a side business selling meat from the cull cattle. The animals that didn't live up to the breed's physical standards were butchered and the meat sold. As Bryan and Cathy's interest in tobacco waned, their cattle businesses grew.

So Bryan did what any ambitious beef farmer does in North America. He called a feed company salesman to ask how to grow a good product: What is considered good beef on the market and how could they produce it? "Grade A, triple A marbled. They defined how good it was by how much fat there was in the meat," said Bryan. "It's a pretty handy game for the feed business. All they need to do is sell you food to make the cows fat—rather than food that keeps them healthy." To grow those grade A, triple A marbled steaks meant a diet of whole corn for the cows, with the choice of adding antibiotics to the feed or even implanting a disc in every animal's ear that slowly released growth hormone. The problem was, Bryan and Cathy themselves didn't want to eat the meat that was produced this way. Cathy in particular was uncomfortable. Bryan remembered: "As a mother she said, I would rather not feed that to our children." After a few years of feeding the cows grain, they started to let them eat grass.

To this day grass is all that the cows eat at Y U Ranch. In the spring, the animals graze on clover, fescue, brome, and orchard grass. When the days heat up in summer, they move to fields seeded in the tall grass prairie species that are indigenous to the area that Bryan also hays to feed the cows in the winter. This is a polyculture blend of three grasses and nine different flowers, a mix he and other farmers have developed with the help of a community-led conservation program called Alternative Land Use Services. Unlike the corn that most beef farmers give their cows, which must be planted every year, Bryan needs to seed his fields only once to feed his animals because his perennial grasses and flowers return year after year. What he plants is also very hardy—Bryan calls them perfectly engineered by nature for climate change. The grasses and flowers in his fields don't need any fertilizer other than the manure the cows distribute, and the plants do their part sinking carbon and even thrive in the heat. The grasses germinate at 22 degrees and are still happy when the temperature climbs above 30; further, because their roots go down three to five metres, they are drought resistant. Bryan never has to buy feed because even in a drought, the grass just keeps growing. "I've cancelled my crop insurance. I don't need it anymore." Talk about building a resilient food system.

The Gilvesys' transition from a tobacco farm to a grass-fed cattle ranch got a push in 2003. The mad cow crisis that was unfolding at the time closed the Canada-US border to live cattle and took away the market for Bryan and Cathy's breeding animals. Since they could no longer sell cows and bulls, they focused on selling meat butchered from their animals that ate a vegetarian diet, free of antibiotics. It was good timing: interest in what they were selling went up as people searched for an alternative to the meat that was implicated in mad cow disease. Today, theirs is a medium-sized farm—225 head of cattle, compared with a feedlot where you could find thousands of animals. And they are an example of how sustainable farming can operate in the developed world—that farming in tune with nature isn't confined to poor countries or the history books. In fact, Bryan is doing better than most conventional farmers because he has created a resilient system, with few inputs, that produces a high-value product and is demonstrating that sustainable can even be better.

Bryan's cattle may, at the scales, yield less meat than cows raised in an industrial feedlot. It takes the calves at Y U Ranch twenty-four months to reach a market weight of 1000 pounds, whereas conventional cattle reach 1400 pounds in fourteen to fifteen months. But Bryan stresses that judging a grass-based operation like his by measuring the weight and age of the calves misses the point. What he and Cathy have built is a system that doesn't rely on fossil fuels like cattle feedlots do. It's a system that works with few inputs. It's a system that stewards the land. And I'd guess that from the way Bryan gets excited when he talks about the living, breathing ecosystem that is his farm, it is his role as custodian of the land that makes him most proud. His farm is not just a business—it's an ecological restoration project. "To say a farm exists outside of nature is a crock," he said. "It's a working landscape. It's not like we have a farm here and nature is over there. If you can get farmers to understand that, you can change the world."

Our tour started in the forest, which is part of the Carolinian zone that stretches from Southern Ontario all the way to the Carolinas in the United Sates. It's an area that is exploding with biodiversity but where environmental damage has been severe. Many native species have almost disappeared, such as the American chestnut, which was logged in the early 1900s because it produced tall, thick trunks that were perfect for the masts of sailing ships. The forest also has been under threat from industrial agriculture and development in general. In the woods on Bryan and Cathy's land, they still have some American chestnuts, great wide trunks standing beside black oak, white oak, and white pine, the tall timbers and the canopy of leaves overhead offering us welcome shade from the sun. "Does this not remind you of a cathedral?" Bryan said.

Their story shows us that if we value farming and food, we can create opportunities for rural development even in a country like the United States. We can slow the flow to the cities of farmers and people living in the country and at the same time preserve the landscape and steward the environment. When we debate about the future of food, we often lose sight of the farmers. But they will play an important role in a sustainable food system by the year 2050.

The industrial food system doesn't need many farmers to function. With a whole lot of capital and the right technology—tractors, sprayers, combines, seeders, harvesters, and more—one person can farm thousands of hectares. In fact, these days you don't even need a human to operate the fertilizer applicator, because a GPS can do it while the farmer sits in the tractor cab, supervising. The government of Canada will even subsidize the cost of that tractor, because it is more environmentally friendly for a GPS to deposit precisely as much fertilizer as is needed in a particular spot than to spread fertilizer scattershot across the field.

In contrast, sustainable agriculture needs people. Lots of people. It's generally agreed that if you take away the technology and the chemicals of conventional agriculture, you are left with more physical work for people to do. According to a 2007 report from the Food and Agriculture Organization, organic farming creates 30 percent more jobs—more if there exists on-farm processing or direct marketing.[24] If you believe that farming is something that we as a species are evolving out of, leaving only the specialized practitioners of agriculture to feed us, this isn't good news. But if our goal is to ensure that every human has enough food to eat, while living within an ecological balance that doesn't endanger the ability of future generations to feed themselves, then the fact that sustainable food systems can involve many people doing many different jobs can be seen as a path to the future.

At the same time that the world is experiencing the largest ever human migration, emptying the countryside and its farms of hundreds of millions of people, there is movement in the opposite direction. Some people are choosing to stay behind to rebuild communities in rural areas. There are also city dwellers moving to the countryside to farm—admittedly a trickle compared to the Niagara Falls moving in the other direction. However, in every populated continent, people like Bryan and Chandrakalabai are actively rebuilding food systems and figuring out how they can produce food, while making themselves an integral part of our future. It is these farmers on whom we depend to grow our food sustainably, to steward the land, and to help with the important job of sequestering carbon.

Jan Douwe van der Ploeg is one of the world's leading academics who looks at these issues. He is a professor of rural studies at Wageningen University in the Netherlands, one of the world's top agricultural schools. Since the 1970s, van der Ploeg has been studying agrarian sociology, trying to understand the changes in farming communities brought on by the industrialization of agriculture. Early in his career, he spent several years in countries such as Colombia where he was involved in creating farmer co-operatives, and later, in Guinea-Bissau, he worked on the reconstruction of rural water supplies. Since then, he has focused on academia, studying primarily what he calls "peasant agriculture" in Europe, Asia, and Latin America. He sees things differently from the conventional perspective found in institutional reports and development agendas, and his viewpoint offers an unexpected interpretation of what's happening in farming communities around the world today. This work has brought van der Ploeg significant notoriety; in peasant studies circles, he is a rock star of sorts.

I'd been trying to get an interview with van der Ploeg for years; he's not partial to talking on the phone, and he was in the Netherlands and I in Canada. It wasn't until I was in Beijing that our paths crossed, and we met one dark evening in the back room of a café on the China Agricultural University campus, where he is also a part-time professor. I soon got the impression that he is a serious man. When we headed across the street for dinner, where we mistakenly ordered snake instead of eggplant from the photos on the menu, he kept a straight face and didn't laugh while the rest of our non-Mandarin-speaking party giggled.

Van der Ploeg's presence in real life matches his academic heft: he is a big man, well over six feet tall, with broad shoulders, big hands, big ears, and a deep, full-bodied voice. He believes that there is a "rupture in the trend to modernize farming." While the mainstream narrative about agriculture may hold that societies around the world are following the drumbeat of modernization and moving away from agrarian societies and towards industrialized modern ones, van der Ploeg sees the opposite. Rather than viewing the small-scale farmers of the world as anachronistic, relics of a past that will soon fade into history, he argues that it is in fact industrial agriculture that is on its death bed. The conventional farmers, who are still producing in the industrial model, are the ones who will soon be squeezed out of the market by an economic model that doesn't pay people properly to grow good food and instead places a downward pressure on their incomes—while at the same time damaging the environment. In a 2010 article, he wrote of the status quo: "Prices are too low, costs are too high, regulatory schemes are too asphyxiating, markets too turbulent and banks too restrictive." He believes the new economy of food will inevitably displace the old.

Wherever he travels, he has noticed the same patterns. Small farmers in China and the Netherlands, Italy and Brazil are refocusing the farm on nature and finding creative ways of making a living in an effort to sustain a rural way of life. He calls these people the "new peasants." To understand who these people are, you must first reject all your preconceptions. Forget the negative associations with the word *peasant*—the small, drafty, dark homes and the starving-thin cow straining at a rope, the misery and the poverty. The new peasants, as van der Ploeg sees them, are their own masters. They are autonomous actors who produce their food on their own land and sell it the way they choose. Sure, these new peasants can also be poor farmers in the developing world with few resources, but the category also includes the middle-class producers—like Bryan—of the developed world. They own tractors and drive trucks and even farm decent-sized tracts of land. But what makes them all peasants—as opposed to what van der Ploeg calls the agricultural entrepreneurs of the industrial food system—is that they are growing and selling in a food network that they are helping to create. A network that permits them to hold on to a certain amount of economic power and in which they place themselves directly in the centre, unlike the farmer's position on the fringes in industrial food.

Hundreds of millions of these small-scale family farms are producing food in developing countries alone. More than half of them are in China and India. In Latin America, about sixteen million small farms produce 77 percent of all beans grown, 51 percent of corn, and

61 percent of potatoes.[25] "I want to stress that peasants are kicking and alive and have a crucial role in our society," said van der Ploeg. "There are many people, particularly women, who say regardless of the experts who say that you can't make a living on a small farm, these women say regardless of everything, we will make a living. There is a lot of doggedness and creativity."

These millions of peasants around the world are represented by an organization called La Via Campesina. The group speaks for the landless, for indigenous communities, for both small and medium-sized family farmers, as well as for migrant agricultural workers and farming women. The name is Spanish for "the way of the peasant," and the peasant's path from the group's vantage point is sustainable small-scale farming. It deeply opposes corporate agriculture and the multinational corporations that dominate the global industrial food supply, and instead promotes local and regional food systems. La Via Campesina was founded in Belgium, has roots in Latin America, and is an umbrella organization for about 150 groups (in seventy countries), including the National Family Farm Coalition in the United States, the Korean Women's Peasant Association, the Landless People's Movement of South Africa, and other coalitions from Brazzaville to Madagascar, the Caribbean and Europe. La Via Campesina is said to represent 200 million rural families worldwide.[26] The idea that draws all these disparate groups of people together is food sovereignty, a concept they popularized.

Food sovereignty is an articulation of the right of people everywhere to choose what kind of food they wish to eat and what kind of agricultural systems they wish to produce it. La Via Campesina believes all people should have access to healthy, culturally appropriate foods that are grown sustainably. The concept of food sovereignty is powerful because of its clarity and its universal appeal. Why shouldn't everyone have the right to choose the food they want to eat? Arguing against the right to choose your food is like arguing against universal suffrage. And food sovereignty as articulated by La Via Campesina provides the same vocabulary—the same framework—to everybody who wants to talk about food and food systems.

I would argue that the simple existence of La Via Campesina is fundamentally profound. The organization is proof that millions and millions of people around the world not only want to support sustainable food systems but would choose to work and live as producers in these systems. When we talk about the inevitability of the disappearance of the family farm, it's hard to argue that these 200 million people represented by one organization are nothing but a blip in the narrative of the vanishing peasant. And these people accomplish things. It has been reported that member organizations were involved in overthrowing national governments in Ecuador and Bolivia. They have also shaped state policy in India. In 2002, a farmer protest in Chhattisgarh, India, stopped the American biotech corporation Syngenta from acquiring varieties of rice seed from an Indian university and taking them back to the United States. These peasants aren't fading into history quite yet.

And people want to farm! Over the last decade there has been a marked rise in interest in farming from a wide range of people. As more and more individuals are attracted to the occupation, the number of small farms in the United States is on the rise. According to the 2007 agricultural census, the number of farms in that country rose by 4 percent over the previous five years—that's a net increase of 75,810 farms. These farms, the census found, were on the whole small farms with younger farmers, and of those that farmed organically, most reported selling their food locally.

All over the world there are stories that speak to this surge in interest in farming. In Japan, it has come from an unlikely demographic and took on some cachet in 2009 when a pop star turned entrepreneur named Shiho Fujita, who had made a fortune selling cosmetics and clothing, rented twenty-four hectares to grow rice. She founded a movement she called Nogyaru (blending the Japanese words *nogyo* for agriculture with *gyaru,* slang for girl) to help reconnect young people to farming. She invited young women, known as gyaru girls for their predilection for dressing in micro-miniskirts and wandering a trendy district of Tokyo, to come and volunteer on her farm. The rice they grew using traditional methods of pest control was sold online. Her website features photos of stylishly dressed young women making peace signs for the camera, surrounded by green stalks of rice. While this example verges on the absurd, it does show that even farming can be sexy, as long as you have the right spokesperson.

Elsewhere, the trend has the potential to be more lasting. In Greece, during the financial crisis in 2012, the *New York Times* reported that young Greeks were leaving the cities, where unemployment levels were soaring, and moving to the countryside to make a future in agriculture. Nikos Garalas and Alexandra Tricha are a part of this fledgling movement. The couple, who are in their mid-thirties, met while working in Athens as farm inspectors for the state but left the city to open a snail farm after they were laid off and the prospect of finding another job seemed remote. "Some of my friends, they did the same," Garalas explained to me over Skype from their rented farmhouse on the island of Chios, where they run their business. "They used to live in big cities like Athens and Salonika, and now it is back to the islands. If they decide to stay in big cities, they will feel the problems of the crisis more deeply than we are."

Garalas was born on Chios, so it made sense for the pair to locate their snail farm in one of the island's greenhouses, abandoned after

European Union subsidies for export agriculture dried up. Inside the glasshouse, they have created a mini ecosystem for the snails, replete with the mastic tree, an indigenous conifer that grows only on their island and will flavour the snail meat. The interest in their farm from the media and other young people has been enormous. The couple has been asked to give lectures and share their story. When I asked them why they felt what they were doing resonated with others, Tricha told me, "People miss hope. There is a hopelessness in Greece. Maybe we represent a little bit of positive thinking."

"If this business works well, we may have our dreams come true," Garalas added.

There is a similar questing for hope in the Pyrenees of Catalonia in northern Spain, where a small organization called Rurbans, working in the area of rural development, opened a shepherd school to train a new generation. Because the average age for mountain shepherds who keep a local breed of sheep called the xisqueta was climbing and their skills at risk of disappearing, the group wanted to facilitate the transfer of the shepherds' knowledge. In their first four years, seventy-two students graduated and they turned away many more applicants. "Most of them have a university education and a strong conviction," explained Vanesa Freixa. The age of the students averages in the early thirties, and most have studied in fields such as agricultural engineering, environmental studies, and social work. The beginner shepherds spend five months in the program, first learning in the classroom and then out on the mountains with the animals. The sheep are kept for both their meat and their wool, which the school's parent organization helps to sell directly to the public through a co-op. They've achieved good results, said Freixa; more than 60 percent of graduates have found jobs as shepherds, travelling up the mountains for the summer months to accompany the elderly shepherds' flocks.

They are able to earn a decent wage because the European Union offers subsidies to mountain shepherds in an effort to preserve this way of life. Now they hope they can transform this old occupation into a viable profession for the twenty-first century.

"The mountains are very hard, there is little land," said Freixa. "The only way to preserve it is with the little farmers. When people visit and see these incredible landscapes, they aren't aware of who is taking care of it. Behind this landscape, and this food, are a lot of little farmers." If the xisquetas did not graze on their mountain pastures, the entire Pyrenees landscape, which has been shaped by human agriculture for millennia, would change as forests reclaimed the grazing lands.

The Rurbans school is one of four shepherd schools in Spain that are offering similar programs in different regions. In the Pyrenees, it's changing the community too. Until 1991, the population was on the decline. However, ever since, the numbers have been increasing as people of Freixa's generation return. "They want to take care of the land," she said. "They want to make their own food and give to society. So they are recovering ancient knowledge and ancient ways of working the land."

This rise in interest in farming is larger than a back-to-the-land movement. The people who want to be farmers aren't necessarily just wanting to escape city life. They are also looking for ways to bring farming *into* the metropolis through urban agriculture. Elaine Howarth is a co-founder of Cultivate Toronto, an organization that has volunteers growing vegetables in borrowed backyards. She says an advantage of being an urban farmer is that you can sleep in, because you don't have to be up with the sun to feed the animals, though "people sometimes look at me funny when I ride the subway with a shovel."

People are dreaming up all kinds of innovative ways to farm in the city, blending architecture and design with farming. In New York, I visited one of the Brooklyn Grange's two urban farms on the half-hectare roof of an old building in the Navy Yard. Ben Flanner, the head farmer and one of the founders, has an industrial engineering degree and worked in business for five years, but he wanted to start a commercial vegetable operation in the city, along with his partners, and run it as a for-profit enterprise. To set up the unusual farm, they laid down the earth on the roof and chose vegetables that could withstand the harsh conditions there—it is windy and gets constant direct sun. The vegetables they produce are sold to restaurants as well as to the public at markets. The project was financed through a combination of loans, grants, private equity, and old-fashioned fundraising. "We make it work by being creative and adhering to basic business principles," Flanner told me when we met on the roof one September afternoon. Volunteers had set up a market stand in the middle of the roof and were selling the vegetables they'd harvested to the public.

In Montreal, since early 2011, Lufa Farms has operated a three-thousand-square-metre hydroponic greenhouse on the top of an office building, where it grows dozens of different kinds of vegetables that it sells to the public through a weekly subscription service. The business has done so well that it is expanding in Montreal as well as looking at starting greenhouse farms in Toronto and in the United States. And Ran Goel left his corporate law job on Wall Street in 2011, when he was thirty-one, and headed to Toronto, where he co-founded Fresh City Farms. Every week, Fresh City delivers a box of organic food to its subscribers that includes the vegetables it grows in its greenhouse and on its urban farm located on borrowed land in a park. The box also includes vegetables from urban farmers Fresh City contracts with who work in backyards across the city. As with many of the new urban agriculture endeavours, Goel sees his business as helping to connect people with their food while reducing the environmental impact of the food system. These entrepreneurs add another dimension to van der Ploeg's new peasant.

The desire to return to the land is not restricted to the West. According to Jan van der Ploeg, who has been studying the case of Brazil, people in that country are leaving cities to farm. "In the 1970s in Brazil, millions of peasants were forcefully removed from their land to make way for coffee plantations," he told me. These are many of the people who have lived in the country's urban slums, the *favelas,* famous for their high crime rates and non-existent public services. Now these people are turning away from this life. "They say we will not accept any more that our children live like human litter," he said. "They prefer to go without the urban TV for a while to create a life with dignity. A dignified environment for their children to grow up." In China, too, migration isn't only in one direction. "If you interview farmers, they will say we don't want our children to be like us. We want them to go to the city and get a job," said Chang Tianle, who used to work for an American agriculture and trade institute in Beijing and now runs an organization in the city that puts on farmers' markets, often in office buildings. "But there are more and more migrant workers returning to the villages. It is almost impossible for migrant workers to have a good life in the city." They go back to where they have a right to the land, to grow food. "It's the social security they return to," she said.

But it is complicated. Everywhere, there are contradictions. As one group moves one way, another moves in the other direction. In China, while peasants migrate back and forth between factory and village, people who have been to university and have had access to a better life in the city are turning to farming.

On the outskirts of Beijing, I visited Little Donkey Farm one cold November morning. The farm is located on fifteen hectares in an area where urban sprawl is quickly taking over arable land. When I visited, a woman named Shi Yan ran the organic farm; it was she who gave it the name, in honour of a droopy-eyed donkey. Since Shi moved there in 2008 to start the farm, a third of the people in the local village have sold their land for development, abandoning farming forever. Their old homes were demolished and they were all relocated to a brand-new apartment complex, where Shi and her husband rented a unit. Instead of fields to grow vegetables, some had brand-new cars they bought with the money they received for their land. Parked below the apartment towers were rows of shiny new vehicles.

Shi Yan grew up in a city in Hebei province, near Beijing. Her mother worked in a dye plant, her father in a power station. The only recent exposure her family had to farming was during the Cultural Revolution, when her mother was sent by the state to work on the land (her father went to be a village projectionist, showing propaganda). Shi decided she wanted to be a farmer when she graduated from high school. She went to university to study agricultural economics, later attending graduate school at Beijing's prestigious Renmin University, where she focused on rural development. It was her professor, Wen Tiejun, who is a powerful proponent of sustainable agriculture in China, who helped her travel to the United States, where she spent six months working on a small vegetable farm in Minnesota. There, she learned how to grow food and how to run a Community Supported Agriculture, or CSA, program.

When she returned to China, Shi Yan launched the country's first CSA on a brand-new government demonstration farm. She and a group of young men transformed the land into vegetable beds; they would later keep chickens and hogs too. The CSA idea was novel in China: members pay a lump sum at the beginning of the season for a weekly allotment of the harvest. They also share in the risk of farming. If the season isn't a good one, the members receive less food. At first, the Beijing CSA attracted 54 members, and then quickly grew to 250 members in one year, with hundreds more on the waiting list. For the farm's ten intern positions, about two hundred students apply. Businessmen visit the farm to see if they can copy what it is doing. The idea has spread. In 2010, Shi and her colleagues estimated that there were already about thirty CSA farms in China—a mere year after she'd brought the model back from the United States. By 2012, she believed that number had grown to at least a hundred.

After a tour of the farm, we ate lunch in her apartment with a coterie of aspiring farmers who were planning to start CSAs of their own. One man, Zhong, who was tall and thin and lived about thirty minutes away, had learned about Little Donkey Farm in the newspaper, then worked there as an intern. Now he was starting a small CSA of his own; he gave me his advertising flyer.

"Our goals are beyond Little Donkey Farm," Shi told me. She and her husband left the farm a while after my visit and started a CSA of their own. "We also want to push the change of our farming ways back to the sustainable way. We often say in China it's hard, but it is also not hard. We have so many years of organic farming and only thirty years of chemical farming. Maybe it is easy for us to turn back."

I asked Shi Yan about the forces pulling people in her country in two directions, to the city and to the farm. She reflected for a moment. "There is an old Chinese saying. *Jie di qi,*" she said. The literal translation is "connect ground energy flow." "That means connected with the soil and the earth. People will not get such a feeling when they are living in a tower block." This makes her hopeful.

In India, too, there is a similar pushing and pulling. Educated urbanites are returning to agroecology and rural life. Among them is Raghavendra Rao, who trained as an engineer and spent a good part of his career working for organizations such as the United Nations and the British Department for International Development on the sustainable agriculture file. In 2002, when he was forty years old, Rao left the capital of Delhi with his wife and daughter to buy land in the southern state of Andhra Pradesh. They opened the Oasis at Kuppam, an organic farm where they grew vegetables to sell in the nearby cities of Bangalore and Chennai. "I had never farmed. I was a first-generation farmer," Rao told me after we met in Bangalore. "I wanted to do something different. I was convinced that organic farming was a viable way to earn a livelihood in a rural area." And that he did. On ten hectares in the arid tropical landscape, he implemented a water-conservation strategy and set up a drip irrigation system, with the help of a government grant. Then, he watched his vegetables thrive without any external inputs. "I used to get amazing yields. They were much more than the yields you could get from conventional agriculture."

He sees now that he was part of an educated demographic of people in India who have land and want to farm in a new way. "A lot of people are getting back to rural areas to try to do something. These include a few romantics," he said, referring to news stories of young IT entrepreneurs who'd made a fortune and wished to leave the city in search of a peaceful farm life. "But generally, the point is to do something for the common good." Today, Rao is the dean of Prakriti Karyashala, "nature workshops" run by an NGO in Andhra Pradesh that teach sustainable farming. In 2010, he was forced to sell his farm for a combination of reasons, including a lack of access to labour. Since the farm was close to Bangalore, a local politician organized daily mass transit so labourers could commute to more lucrative jobs in the booming construction industry there. In Rao's case, manufacturing pulled harder than farming.

All over the world, farming is in direct competition with industrial development and other jobs that society values more than the growing of what it needs to survive. We place a higher worth on making consumer products than food. "In the past twenty years," said Chang Tianle in Beijing, "the best job opportunity for farmers is to work in a factory, in the city, to bring them cash income."

The fact is the vast majority of people are less like Freixa in Spain, Rao in India, and Tricha and Garalas in Greece, returning to the countryside to build sustainable food systems, and more like a man named Umesh I met in Bangalore. While I visited the city, he drove me around. Umesh was a little younger than me at thirty-three, but his thinning hair and downcast face made him feel like my senior. He was tall and very proper, always wearing a white suit shirt, calling me madam, and rushing to open the door for me even though I kept saying it wasn't necessary. He had worked as a driver in Bangalore for four years after having left his village with his wife to work in the city so he could pay off a debt. He found his job through friends and made enough money to save a little. Umesh told me that his father had become ill with thyroid cancer and the family had borrowed the equivalent of about $2000 from the village moneylender to pay for the treatment. As a surety on the sum, they had to give the moneylender a portion of their farmland, where his family grew rice, finger millet, sugar cane, and peanuts. He also had to pay the man interest, which at the village level in India is notoriously exorbitant. "I want to get back as soon as possible because I don't want to live in the city," he said. "Here life is very expensive. We can't afford it. It is better to be a farmer. It is a good job. I enjoy it. In driving, I have to work under someone. When I am a farmer, I follow my own wishes."

Umesh figured it would take him eight years of driving tourists and business people around to repay the sum and head home to take back the farm. If he ever does return. While he yearned for the autonomy farming would offer him, he explained that it was a challenge to make a living growing the way he had done before, and he was doubtful about the potential to make a profit in the future. "In India, the farmers are having so many problems, madam. Every year the price of fertilizer is going higher. We have to buy from a private shop, and in season they increase the price. We have to pay for pesticides and seeds also. And sometimes we don't get the water supply and we don't have electricity to properly run the irrigation pumps. The food prices at the market are so high, but it will not reach the farmers. The middleman takes it. For what we are growing, the prices will be low in the market.

"What you see in Bangalore, thirty to forty years back it was agricultural lands. Now it is a concrete city. The people are not worried about that because everybody wants a luxurious life nowadays. But some people are worried about the future. No one wants to do farming. In ten to twenty years, we will have a food problem. That's why they are worried. Everybody is leaving the farm for the city. Who will grow thc food?"

Indeed. Who *will* grow the food? If we want to build sustainable food systems that can feed people, we need farmers. A paradox of local farms in the West is that their food is often grown by migrant workers who have travelled from far away in search of work. In the United States today, according to the National Center for Farmworker Health, Inc., there are more than three million agricultural workers who depend on seasonal employment, many of whom migrate from state to state, and from farm to farm, picking up work where they can find it. Almost three quarters of these workers were born outside of the United States and come from Mexico as well as Central America—some through official channels, others making the perilous trip into the US across the desert and the Rio Grande. It is estimated that at least 50 percent—and as many as 75 percent—of workers on American farms are undocumented migrants. The workers don't earn much and are often paid by the amount they pick or by the day, or the hour, stooping for long periods in the hot sun, without any job security or health care benefits. These are jobs that Americans don't tend to want. The United Farm Workers of America, a union representing agricultural workers in ten states, responded to critics complaining that migrants from other countries were taking jobs away from citizens by encouraging legal residents to apply for farm jobs. There was no rush to fill positions. The campaign highlighted the obvious fact that agriculture is dependent on immigrant workers who tolerate working conditions and remuneration that people with other options would not.

Farmworkers' rights organizations have brought attention to the often poor and exploitative working conditions on large farms in the industrial food system, and books such as Barry Estabrook's *Tomatoland* have documented how some people are even forced to work without pay on tomato plantations. According to the Coalition of Immokalee Workers in Florida, over the last two decades, there has been at least seven cases of forced labor slavery that have been successfully prosecuted in US courts and more than one thousand people have been freed from slavery. But little attention has been paid to the working conditions on the smaller scale organic farms. When considering how to ensure that an ecologically sustainable food system is also an ethical food system, we must consider the lives of those who work on the kinds of farms we need for the future.

While there are some farms that are leaders in the sustainable food movement and not only pay a living wage to their workers but also provide healthcare coverage, it is possible that the organic kale in your CSA box was picked by a migrant worker who is working in exploitative conditions, said Jessica Culley, a community organizer with the Farmworker Support Committee who works in New Jersey. "As a society in the United States, there is a syndrome of cheap food," she said. And even farms that are trying to be part of a sustainable solution have to work within the system. "They have to base their prices on what people pay at the grocery store." The legions of unpaid interns who volunteer their time on organic farms are another source of inexpensive labor.

Not too far from the Y U Ranch in Norfolk County, Ontario, Jessie Sosnicki and her husband grow vegetables with the help of three workers who come to Canada from Chiapas, Mexico, for several months of the year through a temporary farmworker government program. (The American government runs something similar.) Sosnicki pays her workers $10.25 an hour—that's minimum wage in Ontario. She would like to pay them double but she says the farm can't afford to pay a higher wage because she and husband are trying to make ends meet. She fingers the cheap food in the supermarket that she must compete against. "The more people who keep buying cheap, crappy food," the lower the wage that farm workers are paid. "It's a vicious cycle." It's for this reason that the Agricultural Justice Project has created a fair trade certification program for farms and food businesses in the United States and Canada. By offering verifiable certification to farms and companies that pay their workers a living wage, they hope consumers will chose to buy fair trade and put an end to the cycle.

To encourage new farmers as well as to help people to keep farming—people like Umesh who would prefer to farm but who are drawn to the city because there is a better chance to have an income—farming must pay a living wage. The fact that family farmers are struggling in the United States and India and China is yet more proof of the failings of the industrial food system. We *need* farmers. Particularly family farmers, who provide ecological goods and services to society as stewards of the land. And these small farmers don't need to be poor. "In many situations, it admittedly is a life of misery," Jan van der Ploeg said to me. "But this is not the case for any intrinsic reason. It is because they are plundered. What they have is taken away. In Latin America, they take your water. In China, they take your land for thirty years. In other places, they block the use of local varieties to create the market for GMOs"—genetically modified organisms. "If the neighbourhood landowner diverts the water and you are unable to grow, does it mean that you are a stupid bastard unable to make the land blossom? Or is it that you are subject to an injustice?"

When it comes to the future of farm labour, it's complicated. Although farm work is starting to attract a new generation, it is by no means a universally alluring career path. The reasons for this are obvious: it requires a lot of hard, physical labour, it is often dangerous, and in today's world, the pay is terrible. Bad working conditions and poor money have been pushing people away from the farm for centuries. But we need these farmers more than ever before. The question for the future is, how we can help to make the shift?

"How do we manage the transition to reroute labour away from call centres and service centres and back into the fields?" said geography professor Evan Fraser of the University of Guelph. "We will have to reorient our labour to a more agrarian way of life. If food prices go up, costs of living in the city will also go up. People will increasingly want to reorient themselves to farming. How do we manage the transition to a higher food society?"

In an attempt to progress to a new food system in Italy, a social movement is creating an alternative to the mainstream. They call it Genuino Clandestino, the word *genuino*—which means "genuine"—playing on the idea of the authentic, salt-of-the-earth farmer. So Genuino Clandestino means "genuine clandestine," and it stands for a pan-Italian movement of farmers, consumers, and food artisans who believe in a small-scale sustainable agriculture and food production system that protects the environment—a system where the farmer and the food artisan can make a living from their work by maintaining a close relationship with the consumer. Italy may seem like it already has a well-entrenched local food system. After all, it's a place where small producers have carried on delicious food traditions for centuries—the pastas, the breads, the cheeses, the cured meats, all the delicacies of Italian cuisine. But it too has experienced a transition in food similar to what is happening elsewhere. "When you think of Italy, you think of the tomato. The tomato is a cliché of Italy," said one of the filmmakers in a collective that made a documentary about Genuino Clandestino. "You cannot imagine, though, how many tomatoes here come from China. The tomato you buy in cans—they buy these tomatoes from China because the price is much less."

The film about Genuino Clandestino portrays a system that looks more like the Italy of the imagination. It's a world where farmers earnestly raise their livestock and grow their crops in sync with nature and then proudly sell what they've created to a public that is committed to a food system where the principles of economic fairness and ecological sustainability thrive. On the ground, Genuino Clandestino is a network of regional groups that have each created local answers to the problems of industrial food. "Genuino Clandestino is an anti-brand," explained the filmmaker. (The collective, based in Naples, has its roots in pirate television, and all its members speak under the pseudonym Nicola Angrisano.) "It wants to be an alternative system to the one that is in operation now. And through the actions of the people belonging to the network, they would eventually like to change the current system. They are not resigned to the current system."

In Bologna, the Genuino Clandestino member organization CampiAperti puts on a farmers' market in the city as well as what is called a solidarity buying group, in which consumers get together and purchase in bulk directly from organic farmers. In Rome, the member organization terraTERRA also hosts a farmers' market; in the Umbrian city of Perugia, a market is held in the old town square; and in Naples, twenty or so producers come together through a group called La Ragnatela—the spider's web—to sell their food in a park at the grey edge of legality, hoping the authorities won't shut them down. By linking consumers directly to the producers, said Angrisano, the groups are helping small farmers earn more money. They are creating a market for organic and artisanal products. And they are helping consumers to have a taste of that food sovereignty that Via Campesina advocates for—the right to have a real choice about what to eat and the right to choose food grown in a particular way.

"The idea is to build a relationship between the citizen and the farmer," said Andrea Ferrante, whose family runs an organic farm north of Rome. Ferrante is also involved with the Italian food sovereignty movement and often speaks on behalf of La Via Campesina in Europe. Another example of an alternative food system in Italy that connects consumers to farmers is "social co-operative agriculture." On these farms, the growers are members of a co-operative and work together to produce food that they sell to the consumer. According to Ferrante, this model of farming is used by a good number of organic farms in Italy. One in his area, Cooperativa Agricoltura Nuova, was founded in the 1970s and now employs forty worker members. This type of farm is called a social co-operative because it has an underlying social mission, providing opportunities to members of disadvantaged groups including people with mental health issues, physical disabilities, and even a troubled past in the criminal justice system. There is also an anti-Mafia social co-operative where people grow food on land confiscated from Mafia bosses. "People are extremely happy with the co-operatives," said Ferrante of the consumers who choose to buy the food produced by these co-operatives. "The tomato they grow in the end is the same. It is even organic! So you are not buying it because you have pity, but because inside the tomato is also a social value."

Which really is a metaphor for the alternative food system. These food systems, based on principles of ecology and fairness, offer us more than merely the abundant food the industrial food system churns out. Just as the tomato is more than a tomato, the sustainable local food system is more than merely a way of getting calories into mouths. Not only does this kind of system provide physical nourishment; it also supports farmers who in turn remain in rural areas where they steward the land. We know there are new peasants who want to do this work. Now we must figure out what we can do to support them so we can meet our goals of feeding the world in 2050. It sounds like a feat. But transitions in agriculture as immense as this have happened before.

Back in Norfolk County, Ontario, Bryan ushered me into an old pickup truck to drive me around the rest of the farm as the late-afternoon sun slid down the sky. Amid the tall grasses, teeming with insects and butterflies, the bird songs from the trees at the edge of the woods, it was hard to believe that this farm was not long ago a tobacco monocrop. The transformation of Bryan's fields demonstrates that it is possible to shift from industrial to sustainable agriculture incredibly quickly. He did it in five years. "There are those of us who believe the system can change and we can be the players," said Bryan. "Industrial agriculture puts farmers in the place where people don't think you have any magic." But if people knew about the magic—the skill, the knowledge, and even the luck—it takes to produce food, and the role farmers play as custodians of the environment, then we might want to treat them differently and ensure that the family farmer—the new peasant—has a healthy future. Farmers must be able to make a living growing food.

However, simply paying more at the supermarket checkout won't guarantee the farmer a larger share of the food dollar because, as we've seen, the farmer doesn't fare well within the economic arrangements of the industrial food system. But we *can* do this in local food chains by connecting the grower with the eater, which is the reasoning behind such projects as Genuino Clandestino in Italy and Little Donkey Farm's CSA in Beijing. Bryan, for one, sells his meat primarily to chefs who have a discerning taste in beef and pay for the quality he provides.

We can also improve the income of farmers by acknowledging a role they play that has long been ignored. We can reward farmers for the work they do as environmental stewards. In Norfolk County, Bryan participates in the Alternative Land Use Services program, whereby a not-for-profit conservation organization pays him for what he calls the ecological goods and services he provides such as carbon sequestration and providing habitat for pollinators. (The program has been replicated in five provinces.) Bryan earns $75 an acre a year for every field he plants with tall grass prairie and harvests after the grassland birds have fledged. He is paid $150 an acre for restoring any environmentally sensitive land, such as wetland or woodlot, and then maintaining it. While Bryan isn't going to get rich from the program, it does offer him financial support. Today his farm is abundant in native species of grasses and trees that he and Cathy planted and now steward. It's also a refuge for insects and birds such as the bluebird and the bobolink, a threatened species.

And when we have productive farms on which the people who work the land can make a living, we are more likely to value that farmland. If good agricultural land means good money, then farmers might not be so keen to sell their property to developers wishing to convert fields into subdivisions or quarries. Protecting the soil in which we grow our food is an integral part of assembling a sustainable food system.

Unfortunately, agricultural land is under threat in every part of the world. Could it be that our efforts to build sustainable food systems by 2050 will be thwarted by what's now happening to farmland?

CHAPTER SIX

Land as Good as Gold: Mega-Parks, Mega-Farms, and the Global Rush for Farmland

On my last day in Bidkin, Chandrakalabai came to meet me at the Mahagreen Producers Company office. It was a sunny morning typical of my time in Maharashtra. The air was still and there was yet another blue sky above the dusty, flat fields. In one field, a few hundred metres before the turnoff, several houses were being built on what obviously had been farmland only recently; the building sites looked as if they had hardly been cleared of crops before the work began. All week, I'd been watching the workers' progress, and the bricks and mortar had just about reached the height for a roof. But it seemed no one paid the construction much attention, and when I asked around about what they were building, people shrugged. I guessed they weren't curious because projects like this one were happening all over the region.

Five other women who also sit on the Mahagreen board of directors, all of them farmers, joined Chandrakalabai and me at the office. The women wore saris, bangles, and sandals. They spread a cotton cloth on the concrete veranda outside and we sat together on the ground, cross-legged. They had gathered to tell me about the success they'd achieved in the first year of their company, but the conversation soon turned to the future. News of the government's plan to expropriate a vast, 2400-hectare swath of farmland to build what some call a mega industrial park, an area created for foreign corporations, was causing alarm. This mega-park was conceived as part of the Indian government's infrastructure project known as the Delhi Mumbai Industrial Corridor, launched in 2006 in collaboration with Japan. According to the project's website (delhimumbaiindustrialcorridor.com, of course), the Japanese will provide loans to help the Indians pay the $90 billion necessary to create the infrastructure, which includes nine "mega industrial zones," each more than two hundred square kilometres, a high-speed freight line, and a six-lane highway connecting the two cities, three ports, six airports, a power plant, and various industrial hubs along the corridor. To ease the development pressure on Mumbai, they're planning to build "industrial townships" like the one planned for the Bidkin area. The stated goal of the project is to stimulate local commerce, attract foreign investment, and achieve "sustainable development" by offering manufacturing opportunities in India. The Mahagreen women, however, didn't think the idea sounded very sustainable and were clear about their feelings towards the mega-park. "We don't think the land should be taken," said Shakuntala Mule, who is the board's secretary. "Land is the only asset we have."

There's something sadly ironic about India paving over its productive farmland to make way for manufacturing and industry. At the same time that that these mega-parks are being built, Indian agribusiness companies, along with the government-owned State Trading Corporation that orchestrates imports and exports, are acquiring farmland in foreign states such as Burma, Kenya, and Ethiopia to grow food and cotton for the Indian market. Rather than rely on their own fields to produce pulses and oilseeds at home, they are acquiring distant soils to grow the food the country needs. India is one actor in an international land grab that so far this century has seen governments and private investors purchase millions of hectares of land in more than sixty countries. While the governments are trying to secure food supplies for their people back home, the investors are banking on food crises. They believe the confluence of rising population, urbanization, and the effect of climate change on agriculture will mean that investments in food, and the land that grows it, will pay off—big time. According to the World Bank, fifty-six million hectares worldwide were leased or sold between 2008 and 2009, and estimates from other organizations reckon even more. HighQuest Partners, an American strategy consulting firm that helps financial investors and businesses decide where to put their money in agribusiness, gauges that between $15 and $45 billion has already been invested in farmland, a number it says will soon triple. The competition for farmland is only going to become more intense.

But if we know that sustainable agriculture in a local food economy helps us to achieve our goals of building resilient food systems that don't damage the environment and that improve communities—particularly in less well-off countries in Africa, as well as in India—then we need to ensure that there is enough farmland to support this. To feed the world under the pressures of climate change, we need farmland near urban markets that serve the people nearby. We also need to ensure that those farmers, who will best help us to achieve our goals, have tenure over this land.

Unfortunately, though, the recent spate of land grabs is moving us in the opposite direction. These land grabs are transforming small farms, forests, wetlands, and other natural areas into large industrial farms to grow cash crops such as biofuels and commodities for far-off markets. The integrity of farmland around the world is one of the most important issues for the future of food. To achieve our first goal of sustainable agriculture in the 2020s, we urgently need to start protecting the farmland.

At my meeting that morning with the directors of the Mahagreen company, I sat across from Yamunabhai Rathod. She is the director of the producer company and, just like the other women there, she is an organic farmer. Yamunabhai looked to be in her late forties and wore a lime-green sari. On her face, she had traditional tattoo markings in blue-green ink; she also wore an expression of deep sadness. It had taken her more than a decade to convert her four hectares to thriving organic agriculture, and only days earlier she had learned through the newspaper that the Bidkin extension of the mega industrial park in Maharashtra would take over her land. "I can't sleep," she said. "We're worrying every night about it. The whole village didn't want to sell. What will we do if there is no farm? Some of our village went to the chief minister to tell him we don't want to lose our land. He said he would do something about it, but now the government has given permission to take away our land."

The scenario has been repeated many times around the world. Big interests and big money arrive in a rural area and the small farmers—or pastoralists, hunters, or nomadic people—are told to get out. Rarely are people compensated, and if they do receive anything, it is usually not a fair settlement. Instead, they are forced to move on and make way for industrial farming operations such as oil palm plantations, sugarcane for biofuel production, or food-producing farms. While the land doesn't always remain in agriculture—other industries are also involved in acquiring land to be used for logging, mining, and, in the case of Yamunabhai, industrial development—according to a 2009 report co-authored by the International Institute for Environment and Development and two other international organizations, the majority of land is being used to grow food.[27] It is, after all, according to an Oxfam report on the subject, the prime farmland that interests investors.[28] And local governments are often helping to clear small farmers and peasants to help smooth the transition from small landholders to industrial farms.

The extent of the investments in foreign farmland is dizzying. The Chinese are growing rice in Cameroon. The Saudis are producing food in Ethiopia and Argentina. Canadians are in Kenya, and a Dutch company is in Sierra Leone. The Vietnamese are starting rice and rubber plantations in Sierra Leone too, and Kuwaitis and Qataris are leasing swaths of Khmer paddy fields in Cambodia. According to the *Korean Times,* eighty-five Korean corporations, including Daewoo, have bought farmland in twenty countries, among them Cambodia, Russia, and Indonesia. And in 2008, Goldman Sachs in the United States purchased Chinese poultry farms. More than 40 percent of farmland in Laos is controlled by foreigners involved in agribusiness. Thai companies are looking to produce cassava and palm oil there, and Mongolia and Kuwait have plans to grow rice there.[29] *National Geographic* reported in 2012 that in a period of five years, the government of Liberia had sold at least a third of the state's land. One of the big investors in farmland is pension funds. The Teachers Insurance and Annuity Association of America, which manages more than $400 billion in pension funds for teachers and academics in the United States, owns more than six hundred thousand hectares of farmland at home as well as in Romania, Poland, Brazil, and Australia.

No one is immune. One financial organization that invests in farmland on behalf of pension funds, called Hancock Agricultural Investment group, manages almost 300,000 acres of prime farmland across the country, including in California, the Midwest, and the Mississippi Delta, as well as in Australia and Canada. In that country, where laws in many provinces prevent foreign ownership of farmland, investors are nevertheless itching to buy in, particularly in Quebec and Prairie provinces. "The interest we're seeing has gone exponential,"

Stephen Johnston of Agcapita told me. His firm is one of a number of private companies in Canada that are creating farmland investment funds and investing on behalf of high-net-worth individuals there. If the law allowed them to, they could find many more investors from other countries. That's because investors everywhere fear being left out. This surge in interest has created the feel of a Wild West gold rush. Which is why people are calling this a land rush.

It wasn't long ago that farmland and agriculture were of no interest to investors, even while development agencies were calling for more financing in the sector, particularly in Africa, in the hope that productivity might improve. For decades, the private sector had looked to invest elsewhere while governments around the world, particularly in developing countries, divested themselves from any state involvement in the sector. But everything changed after the two global crises of 2008, first in food, then in finance. When commodity prices started their hike up to record levels, and the repercussions began to be felt around the world, attention suddenly focused on the issue of our future food supply. At the same time, the world's financial systems were collapsing, triggered in part by the sub-prime mortgage scandals in the United States.

What happened next demonstrated how the food we eat had become intertwined with international financial markets and investments, with negative results. Private equity funds, pension funds, and hedge funds, with the help of their own governments and loans from the World Bank and other organizations, began to make agricultural land and factory farming a strategic asset. If the price of food was going up, it would follow that the land that produces that food would be worth more too.

Traders began investing in agricultural commodities even before the Chicago Board of Trade was created in the late 1800s. The idea was to allow farmers to lock in to a sale price—a futures price—so they would know in advance how much they would earn from their work. The system also allowed producers who relied on these agricultural products for their businesses to guarantee themselves a supply as well as a price for the commodities they needed. These investors are called physical hedgers because they hedge their bets against rising (or dropping) prices. They lock in to a contracted amount of money per unit of food as little as days or as much as months or even years ahead of the actual sale of a good. As long as these are the folks who are buying futures in such things as wheat and pork bellies, then the commodities market reflects the realities of supply and demand for a particular food at that point in time. If many buyers want pork bellies, the price goes up. If farmers have a bumper crop of corn, flooding the market with supply, the price of a corn contract will be lower.

In the 1930s, the US Congress created a system to prevent speculative price bubbles of these commodities. They put limits on the ability of another kind of investor from getting involved in the market. Whereas farmers and producers are physical hedgers—people who want to buy and sell food—a speculator is someone who only wants to make money off the trade of these commodities and has no interest in ever delivering or receiving a shipment of, say, corn. A state-regulated system was similarly implemented in other commodity exchanges around the world such as London, Mumbai, Tokyo, and Singapore.

This is more or less how agricultural exchanges worked until the 1990s. Then, lobbying by the financial sector slowly undid government regulations, and the market opened to more speculators, rather than the physical hedgers. Between 2002 and 2008, more and more financial investors got into commodities (the dot-com market had crashed, after all, and they needed somewhere new to invest), pushing the price of wheat and rice and cotton up and up and up until it reached its height in 2008. When the boom in agricultural commodities was at its height, one hedge fund manager, Michael Masters, estimated in his testimony to Congress that on the American commodities exchanges, index speculators, rather than physical hedgers, owned so many wheat futures that they could "supply every American citizen with all the bread, pasta, and baked goods they can eat for the next two years." The speculators might have owned contracts, but they certainly didn't want to ever buy the actual wheat. They only wanted to turn a profit from buying and selling those contracts. The futures market was so hot that it was driving up the current price of food too—in investor-speak, the market was "contango." The cost of a wheat future was higher than the current price of wheat. What this translated to in the supermarkets of the world was that bread and onions and corn cost more than they had only months earlier. There is some debate among economists about whether speculation did in fact push up commodity prices in 2008, but institutions such as the United Nations have concluded that this kind of investment did indeed play a central role. Olivier De Schutter, the UN's special rapporteur on the right to food, wrote in his review of the crisis: "The 2008 food price crisis arose because a deeply flawed global financial system exacerbated the impacts of supply and demand movements in food communities."

What does all this have to do with small farmers like Yamunabhai? Well, the soaring commodity prices were all it took to turn the agricultural land where those commodities sprout into a hot-ticket investment. The idea that agriculture was a solid investment was further entrenched when scientists and international organizations began ringing the alarm about our species' growing need for more food. Agriculture could in fact pay decent returns! A future of agricultural land scarcity combined with booming population meant this area could pay off in the long run too. When the US housing market began to crash in 2007 and the world's financial systems teetered on the brink of total disaster, investors wanting to diversify their portfolios moved further into agriculture. For some agribusiness companies, the desire to own farmland is a move towards complete vertical integration, whereby a food-producing corporation owns every step of production, from the field all the way to the final product. (This is advantageous to the company because it protects it from price surges and cuts out the middleman.) For others, growing food in the short term is a way to make a profit while at the same time demonstrating the fertility of the land and improving its value for future sales. In Canada, where no data are collected on the number of hectares controlled for investment purposes, the price of renting from the investment firm Agcapita an average eight-hundred-hectare farm in the Saskatchewan breadbasket was $90,000 a year, a sum paid upfront to the investment fund by the leasing farmer; prime farmland is worth even more. Before purchasing a property, the company does its due diligence, sending out agronomists to investigate soil fertility and other features to ensure that the land can attract a farmer willing to pay this high rent. Agcapita expects that as land prices in the province continue to climb, the amount it can collect each year from renters will go up too. Marvin Painter, who is a business professor at the University of Saskatchewan, suggested that the rent, in combination with the predicted capital gain from a future land sale, works out to be worth only slightly below the value of a blue-chip stock. And the rents changed in Saskatchewan seem paltry in comparison to Iowa, where farmland rents are more than six times higher.

In countries like Canada or the United States, this type of investment is rooted in free market principles. Individuals who own property have the right to sell it, and those with money are freely able to invest their funds in land and rent it out as they see fit. However, in practical terms, what happens when a big company comes to town and buys a lot of farmland is that the local community loses control over its natural resources, particularly of the water systems that are the quintessential shared public good. And this tends to worry locals, no matter where they live. In 2010, when reports began circulating in Quebec that Chinese interests were looking to buy millions of dollars' worth of farmland in the province, people were outraged. A Desjardins Group study paper of the foreign purchase of farmland in the province concluded that people's strong reaction to the prospect of foreign ownership demonstrated how important the question of who owns land is to the public. In Florida, when former auto parts magnate Frank Stronach, who used to run Magna International, bought nearly thirty thousand hectares of prime farmland in Marion County, locals started to worry. Stronach planned to open a massive grass-fed beef ranch, with more than thirty thousand head of cattle and its own abattoir and biomass power plant. According to the local media, some nearby residents hoped the farm would bring jobs, and politicians were grateful for the business. But others worried about the amount of water the farm planned to draw. Stronach had asked the local authorities for permission to drill about 130 wells to pump out more than thirteen million gallons of water a day from the aquifer to irrigate the grasslands where his cows would graze. This did not sit well with neighbours.

In developing countries, the arguments in support of investment in agricultural land take on a different tone. In Africa, where most of these land deals are being made, there often doesn't exist the same concept of private land rights that we have in the West. According to the World Bank, less than 10 percent of all land on the continent is held in formal land tenure; the rest operates with customary land tenure, in which tradition is repeated so many times, over many years, that it takes on the force of law within a society. In some cases, such as in Ethiopia and Tanzania, land has been nationalized and the state holds title. This is why it is typically the government, rather than private landowners, that acts as the seller in these deals. A government sells or leases its agricultural land to a foreign entity because it hopes that investment will stimulate GDP growth and even increase government revenues through taxation. Often investors promise jobs, and some land deals include building infrastructure projects. For example, in exchange for a lease of forty thousand hectares in the fertile Tana River Delta in Kenya, the government of Qatar offered the poorer state several billion dollars in loans to build a deep-sea port on the island of Lamu.

But Oxfam calls these deals "development in reverse." The international organization, which has been tracking these land deals and examining their implications for communities in the developing world, states in its report that these transactions will "do more harm than good." Already, plenty of proof demonstrates that these land grabs stand in the way of sustainable food systems. The report put out by the International Institute for Environment and Development states that because most of the land in question is home to farmers, herders, foragers, and hunters, the deals could mean a loss of livelihood to large numbers of people. Buried in the footnotes of the World Bank report "Rising Global Interest in Farmland" is the story of how farmers in the Democratic Republic of Congo were displaced after a foreign buyer acquired their land. Locals were left paying guards at a national park to allow them to grow crops there.

The people purchasing the land aren't necessarily foreigners looking for a good investment. Local elites are buying it too. And not everything is above board. Transparency International, a worldwide coalition of people fighting corruption, has found land management and ownership issues to be particularly susceptible to corruption. For example, in 2012 in Sierra Leone, local chiefs countered the will of the people and leased prime farmland. A Belgian company called Socfin, its subsidiaries already controlling plantations in Cambodia, Liberia, and Cameroon, took out a fifty-year lease on sixty-five hundred hectares to plant oil palms and rubber trees, and was seeking to acquire more. The first part of the deal is worth $112 million. When people protested and set up roadblocks, they were arrested and charged with unlawful assembly and public disorder. According to the NGO Green Scenery, which works in the area, locals have complained of being harassed by armed police and there are even reports of the chiefs themselves threatening young people who were vocally against the project.

Felix Horne is a consultant for organizations such as Human Rights Watch and the California-based think tank the Oakland Institute. He has travelled to Ethiopia and Zambia to research land grabs, but it hasn't been an easy job, Horne told me when safely back in Canada where he lives when he's not travelling. Because authorities generally don't like foreigners nosing around where large swaths of land are being handed over to multinational corporations or other governments, Horne had to do his research incognito, taking on the persona of a backpacker. "That comes easily to me," he joked. He spent many months on the continent, travelling to remote locations where the land was to be transformed into large industrial farms, replete with worker campsites and airstrips. One area he visited was the Omo Valley in the south of Ethiopia, where the government has leased out large tracts for sorghum, maize, and oil palm plantations and is also clearing land to run its own farms. "This is the Africa of *National Geographic,*" he told me. "This is where people wear body paint, traditional clothing, if any, and huge body piercings—with an AK-47 slung over their shoulder. It is the antithesis of Western society." These people he described are pastoralists who live with their livestock and who practise swidden agriculture, an ancient agricultural technique that involves cultivating the land in one place for a time and then moving on to clear another location, leaving the fertility of the soil to regenerate on its own. "People are angry. People are very angry," said Horne. "There is no one at the local level deciding whether this is good for them. There is no opportunity to say no. There is no appeal. They usually find out it is happening when the bulldozer shows up."

When their land is taken, they lose more than their livelihood. "The connection to the land is extremely important," he said. "It's about identity." The depth of this connection is such that when people who have moved from another area, Gambella, where there have been similar land grabs, die abroad, their bodies are flown back to Ethiopia, where they are buried. "That's even being cleared—those gravesites," explained Horne. "Land, like in many places, is everything. It's not just where your food comes from. It is part of your culture, your identity."

The question is whether the greater gains offered by the infusion of foreign money into poor areas outweigh the local negative consequences. Unfortunately, if you look at the details, as Horne did, these local impacts are severe. Although technology and jobs are often promised to these communities, rarely—if ever—do these benefits materialize. "It was very difficult for local people to get jobs," said Horne. "We talked to local investors who said, 'Look at them, they are malnourished and they can't work hard. Why would we hire them?'" Instead, these companies employ people who come in from the cities—in the case of Gambella, it is people from the highlands who belong to different ethnic groups, and this feeds into tensions between people. In 2012, several men from Gambella made their way into a rice-producing agribusiness compound and shot all the foreigners they found. Human Rights Watch reported that in an act of retribution, the Ethiopian military went into the local community and arbitrarily arrested and raped "scores" of villagers. "I hear testimony regularly from people who say they will declare war on Ethiopia. It is a disaster waiting to happen," said Horne. "We are setting up the potential for larger conflict and marginalization for the poorest people on the planet." In the Omo Valley, in the Gambella, and elsewhere too.

When peasants have protested being removed from their land, they have put both their lives and their livelihoods at risk. In Guatemala, small farmers in the Polochic Valley were evicted when a sugar cane refinery producing ethanol arrived in the area. The farmers were forced to move to the steep mountainsides to try to grow their crops, and when they attempted to return to the valley after the refinery pulled out of the region and put the land up for auction, a private security unit moved in and evicted them once more. There have been murders too. In January of 2012, Matías Valle Cárdenas, who was a leader in the peasant movement in Honduras, was murdered on his way to work one morning after protesting the presence of another oil palm company. This man was the latest in a string of targeted murders in the country, and the BBC reported that between 2009 and 2011 about thirty-six people were killed in clashes over land ownership in Honduras.

There are so many similar stories: in Vietnam, a lawyer named Tran Quoc Hien, who was helping farmers through the United Workers-Farmers Organization, was jailed for five years and then exiled after his release. In the Philippines, dozens of people who have stood up for the land rights of small farmers have been assassinated. One Mexican extension worker told me that armed men had visited his home to tell him to stop his work promoting sustainability in his community.

The motivation behind selling, taking, leasing, or appropriating the land that belongs to small farmers is the same everywhere. People want to make money, and the best way to do that is through feeding industry and the world's economic system, not humans.

Wherever I travelled when researching this book, people told me a similar story: in their countries, many believed they didn't need to protect their own farms and preserve the soil because someone else in some other country that had a comparative advantage in farming would grow food that they could then import. In a Beijing Starbucks, Chang Tianle, who founded an organization that runs farmers' markets in the city, explained the politics of land in her country to me. "It has always been a debate in China among the leftists and the rightists," she said. "They say, we don't need to keep our land. If we are good at manufacturing, we'll build factories and we'll import food." In India, people told me the same thing. But if everybody believes someone else will produce food while they fuel their economies with other industries, then where is all the food going to come from?

We can't simply go out and find more farmland. In China, most of the geography is uninhabitable desert, as well as glaciers and mountains, meaning 60 percent of the population lives on less than one-third of the land mass. In the basins and plains, the country has the fewest farms per capita of any country in the world, and feeds 20 percent of the world's population on a mere 7 percent of its farmland.[30] Not only is there not an unlimited amount of land to expand into, but the health of the planet depends on the biodiversity that exists in the forests and wetlands and mountainous areas we haven't turned over to agriculture. Also, if we were to cut down more forests to make room to grow our food, we would hasten climate change because trees sequester carbon, which is released into the atmosphere when they are felled. We must keep the forests as forests and the farmland as farmland.

To ensure that we have enough land to grow the food we require, small farmers and family farmers need to retain control of their property. If we sit back and watch as millions of small farms are transformed into biofuel plantations, cash crops, housing subdivisions, and industrial plazas, the new peasants will lose their ability to grow food and we will all risk losing our food sovereignty, our right to choose the kind of food we wish to eat and the right to choose how we would like this food to be produced. In developing countries, the state may be selling off land because the elites are corrupt and are trying to profit from an easy opportunity. But deals are also made because these countries are poor. They lack basic infrastructure, have flagging economies, and appear to have few other options to improve conditions. These governments are placed in a difficult situation. They are offered a false choice—a choice with no right answer, a choice no one should be forced to make. They must decide between selling their property to wealthy foreigners to hopefully make some money, create jobs, and build infrastructure or allowing small farmers in their countries, who live in poverty, away from the benefits of modern society, to continue to eke out a subsistence living at best.

In the West, farmers are keen to sell their acreages to corporate buyers such as developers who will pay millions for the opportunity to turn a farm into a subdivision because, like the rest of us, they want to retire comfortably. (And municipalities are keen on the increased property tax revenue.) This situation is even acknowledged in a report on land acquisitions written by the Desjardins Group. It wrote that farmers sell their properties because "they are being squeezed by climbing operating costs." In the province of Quebec, 30 percent of farms can't cover their expenses.[31] "Beyond all these considerations, farmers are increasingly feeling isolated," the report reads. "Farming is less socially acceptable, for lack of a better term. Urban expansion to the suburbs and rural regions may have brought urban and rural dwellers in closer proximity but it has also made it more difficult to run a farming operation. For all these reasons, plus the financial problems, some farmers are thinking about throwing in the towel." Such are the words of a bank.

Solutions to stop the loss of farmland are coming from the grassroots. In France, people are trying to tackle this issue with an innovative plan. "France is divided into ninety-six territorial *départements*," explained Valérie Rosenwald, coordinator of Terre de Liens, an organization that works to protect agricultural land from urbanization while matching organic farmers with property. "We are losing farmland at the rate of one *département* every seven years. There's a moment when you realize that you have to do something about this." So her organization thought up a creative way to protect this farmland while at the same time helping new farmers get into agriculture. It is creating a living land bank by both accepting gifts of acreages to be protected in trust, in perpetuity, as well as using donations of money to maintain its properties. It raises the funds for the project through an innovative private loan system, borrowing interest-free from individuals. In both cases, the organization retains the title to the land and offers it on a long-term lease to a new farmer. "The idea is to protect the earth and to protect food production as a common good," said Rosenwald. The organization was founded in 2003 and already protects twenty-eight hundred hectares from development.

So far, more than nine thousand people each have loaned on average about 2000 euros—which is more than $2500—though some loans are as little as a hundred euros while others as much as a hundred thousand. "It's a way for people to connect to the earth," explained Rosenwald of people's motivation to lend the money with no tangible returns or dividends. In the future, when someone wishes to collect the principal, the organization will make a modest adjustment for inflation but essentially give back the original sum. The money raised as well as the gifts of land now support about 150 farms where 270 adults work producing vegetables, meat, cheese, grains, and wine, practising sustainable agriculture that Rosenwald calls "peasant agriculture," harking back to Jan van der Ploeg's new peasants. The organization also owns forests and other areas that aren't used for agricultural production but are now protected from development.

In the developing world, people have a different approach to preserving farmland. The Indian city of Bangalore is known as the Silicon Valley of India because it is a centre for the country's booming IT sector. Bangalore is also known as the Garden City because of its trees and parks. However, it is debatable now whether it is still worthy of the nickname. In the last decade the city has not only been built up, destroying a lot of the greenery it was known for, but its borders have also expanded. As is the case with all growing cities, the farmland at the outskirts is being bought up and turned over to housing and industry. The small agrarian-based villages that used to ring the city are disappearing. This is why the town of Magadi, about twenty kilometres outside Bangalore, was chosen for a pilot project designed to protect the land by helping small farmers earn a good income from producing food there.

Magadi is a small town with the feel of a suburb. It was more built up than the other villages I visited in India, and there were obvious signs of improving prosperity. Motor scooters were parked out front of concrete homes, many of which were two storeys tall, and small silk-reeling enterprises were being run out of some of the houses, with noisy machines pulling the threads from silkworm cocoons. But many in the village still relied on farming, and the conditions were right for it. The soil, people explained to me, was a sandy loam, and there's a fair climate and access to irrigation water. Whereas in the rural areas such as where Chandrakalabai lives, the farmers' fields are typically located on the outskirts of the village, in Magadi the farmers I visited grow in what amounted to their backyards.

There, I saw how people squeezed as much food production as they could from their small spaces. At one home, field beans grew along a railing reaching all the way up to the second floor, and at another, the only room a pumpkin could find to spread its leaves was over the top of a large rock. People grew cherry tomatoes, the fragrant herb curry leaf, peas, and eggplant—the climate was so perfect for eggplant that one stalk provided fruit for four months. In their gardens, people also crammed in vermicompost sacks they shaded with the leaves of the coconut palm so they could fertilize their crops naturally, and there were cows and chickens too. The project, now run by an organization called the DHAN Foundation, was designed to promote peri-urban agriculture so that people can value their land as food-producing and maintain it. The foundation teaches agroecology in a farm school to improve yields and has set up a farmers' co-operative to help them market their produce and retain a good profit margin. Whereas a middleman used to bring the vegetables to market and take a cut, these days the co-operative organizes the selling. As a consequence, the villagers have seen their incomes rise by 15 to 20 percent. More than 150 farmers were participating when I was there, and the DHAN Foundation had hopes of opening its own supermarket one day to showcase the local food. It had achieved success so far, but it remained an open question whether what it was doing would be enough to preserve the land and stop the spread of the city. The expanding metropolis is hard to stop.

Civil society so far has played the most significant role in building sustainable food systems, with little involvement from the state or the corporate sector. Considering the lack of resources available to organizations, not-for-profits, and charities compared to what governments and corporations have at their disposal, it's astounding what they've done. But now it is time for governments to play a bigger role. The spread of the city and industrial sprawl is so hard to stop because it is a move against the status quo. To improve chances of success, farmland preservation needs to become the new status quo, and this shift requires public policy. Particularly when it comes to protecting farmland, governments can play a starring role.

One idea is to preserve farmland from urban expansion with a greenbelt. A greenbelt is a large swath of undeveloped, open land that surrounds a city, usually made up of both public and privately owned land, and its purpose is to preserve the natural and agrarian landscapes. In these greenbelts, there may be towns and villages as well as farms, and forests and other natural landscapes. There are many examples around the world. A total of 13 percent of England is protected by greenbelts around more than a dozen cities, including one around London, which was created in 1938. In the Netherlands, open space and agricultural lands are protected near Amsterdam and Rotterdam by what is called the Groene Hart, or Green Heart. After the fall of the Berlin Wall, rather than develop the no man's land that had sliced Germany into East and West, the decision was made to spare the land and preserve the habitat and the ecological goods and services the area provides. These areas complement the European Green Belt, a collaboration between twenty-four governments that stretches from the northern tip of Finland, through eastern European countries such as Estonia, down through Germany, the Czech Republic, and Hungary all the way south to Turkey and Greece. The metropolitan area of São Paulo, with its bulging population of more than twenty million, created the São Paulo City Green Belt Biosphere Reserve. In the Canadian province of British Columbia, the Agricultural Land Reserve has kept around 4.7 million hectares of private and public land from the heavy development pressures in the south of the province, and in the province of Ontario, the Greenbelt that was founded in 2002 protects 730,000 hectares of some of the most fertile farmland in the world.

Society benefits tremendously, not just from food production that happens on the land but also from the natural goods and services. Greenbelts are natural life support systems. So we can support greenbelts by buying the food grown in them as well as making sure there is a market for that produce. This has been achieved through government procurement policies that stipulate how much food universities, hospitals, prisons, and other public institutions must purchase from nearby farmers.

One problem with greenbelts, though, is that they can be subject to the whims of local politicians. For example, in British Columbia, thousands of hectares have been taken out of the Agricultural Land Reserve by municipal governments wanting to increase their tax base or gain favour with developers, though there are still millions of hectares that remain protected, ensuring that local farmers can stay in business. Greenbelts can also cause leapfrog development, in which a city's bedroom communities are simply pushed farther away by the protected band, as was the case around the Canadian capital city of Ottawa. And farmers hoping to sell their farms to finance their retirements frequently protest against greenbelts because they reduce a farm's property value.

But although there are many examples of greenbelts, there are not enough. If we are to save our farmland, governments everywhere need to step up.

It's easy to imagine what the farmland where Yamunabhai grows her food would look like if the industrial mega-park is built. You have only to look at the industrial sprawl around the city of Aurangabad to picture it. The area is one of India's centres for the chemical and plastics industry as well as seed production, and fields have been replaced by sprawling concrete buildings surrounded by high fences. And all around those fences are shantytowns where workers make their homes out of corrugated aluminum and scavenged Styrofoam.

Some of the residents in the shacks are likely those who used to farm the land beneath them. At the meeting with the Mahagreen Producer Company board, when we were all sitting together on the veranda outside, Navnath Dhakane, the CEO, looked at Yamunabhai and put it bluntly. "After industry comes, she will be landless," he said. "They have to leave their own village, their own neighbours. If they are lucky they will go from owning ten acres of land to two acres of land." He spoke to me in English. Yamunabhai couldn't understand his words, but she began to cry and wiped the tears with the corner of her sari.

"The people say we are going to get jobs," he went on. "But they don't hire local people. Management hires outsiders because the local people don't have skills. They are farmers." Then he told me of one village, Chitegaon, that has been swallowed by Aurangabad's industrial sprawl. "Near one company, dirty smells are coming. The groundwater is totally polluted. Nearby in the villages, no crop will grow up. The soil is totally damaged from the pollution from the factories, the wastewater from the factories. One man took the water from there and went to do his toilet. Three days later, his backside had developed rashes. The women were talking about it in a farmers' meeting." Even if this last story, spreading from village to village, was apocryphal, it speaks to the anxiety that local people have about what is happening to their land.

"I think there will be a revolution here," said Joy Daniel, the head of the Institute for Integrated Rural Development. "A single cause can bring the masses together." They'd already witnessed people in Maharashtra coming together earlier that year under the leadership of a man known as Anna Hazare to protest political corruption. "Corruption affects few people compared to the number of those who are affected by injustices in farming," said Daniel. "The price of food makes even those in the city affected." And according to the local media, the land won't go without a fight. In one article, published at the end of 2011 by the Lokmat News Service, about a public meeting between villagers, the local politician, and a land collector, it is obvious that a life without land isn't what people want. The article describes how the land collector was trying to persuade the farmers to sell by promising jobs and the possibility of one day regaining some property. "However, farmers did not budge," wrote the reporter. "They firmly told the collector that . . . they would not like to part with their precious land."

The international farmland grab raises the stakes and increases the challenge in our pursuit of a sustainable food system. It is a daunting hurdle. But while those interested in investing in agriculture may have money and might, farmers and their supporters have conviction. As one of the women in Bidkin told me, "Land is our only real asset." She meant that for the peasant farmers near Aurangabad, who depend on what they produce from the earth to feed their families, land is all they have because without it they can't survive. Yet her statement applies to the rest of us too. Whether we live in cities or on farms, land is what supports us all.

PART TWO

TARGET 2030:

SEEDS

CHAPTER SEVEN

Two Thousand Years of Rice: What China Knows That We Don't

When the morning fog lifted, the Yuanyang rice terraces appeared. The misty curtain rose to reveal the paddies that farmers had built into the steep mountains in China's southernmost state of Yunnan more than a thousand years ago. At dawn, the fog had been so thick you couldn't see past the shadowy green outline of the trees at the edge of town. I had to wait for the sun to burn off the fog to see one of the few functioning rice terrace systems left in the world.

On a map, the old town of Yuanyang is only a fingerprint away from the Vietnam-China border—less than a hundred kilometres—but hard to reach. The closest major city, Kunming, which is the capital of Yunnan, is at least a seven-hour drive away, and the trip is a challenging one. The roads in these parts are more like snakes, twisting around the mountains, squeezing between rice paddies and small farmhouses, often narrowing to a dirt track. That's when there *is* a road. Even though China is rapidly building new thoroughfares, with freeways stretching across the country for the first time in its history, there are still many villages in the state of Yunnan that can be reached only by foot or, possibly, motor scooter. It's this isolation that has helped to preserve the culture of rice that persists here today.

Like the rice terraces, the old town of Yuanyang is built into the side of the mountain. The town has been known officially as Xinjiezhen ever since the local government relocated the capital to the river valley in a move to boost its economic prospects, but I was told people still call it Yuanyang. The houses are built close together, clustered on the incline, separated by narrow streets and pathways with steep staircases connecting one level to the next. The evening I arrived, a traffic accident blocked the road and cars and trucks were backed up for at least a kilometre, no one able to move until the debris was cleared. The sun set quickly, and in town, the bright yellow lights from the shops lining the roads lit the way to the hotel. The terraces would have to wait until morning.

As the sun rose the next day, the town awakened. The roosters that had been calling all night reached a fevered pitch and were joined by even more. In the half-light of dawn, children began to arrive at the school and the drone of old diesel trucks, motor scooters, and three-wheeled motorcycles grew louder. All around were signs that rice was at the heart of this mountain town. A young woman wearing a traditional hand-embroidered blouse and pants stood in front of a small shop on the main road, stirring a giant pot of steaming rice. At a canteen, people lined up at a counter for rice noodles in broth. On the main road, three small trucks passed, piled with the dried stalks of the rice plants that would be used for animal feed and bedding. Outside a seed store, metal rice threshers were lined up for sale.

Life in Yuanyang County has long been built around rice. The grain has been farmed here since the Tang dynasty (618–907), when people belonging to the Hani ethnic group likely migrated from the north and began to sculpt the mountainsides into terraces. To transform steep terrain into fertile ground required a lot of human thought and invention. By harnessing nature's systems, the Hani created what is called an agroecological landscape. They diverted the water that drains from the forests at the top of the mountains through irrigation channels and along the dykes and banks that interweave the paddies below.

The system is ingenious. Every aspect of the landscape was planned by these ancient architects. They chose to preserve the forest at the top of the mountains because that's where rain collects. They built their houses below the forests, thus protecting the trees, and conveniently locating their homes at a high altitude, where it's cool and comfortable to live. They then positioned the rice terraces farther down the hillside because that's where the temperature heats up, making the area more suited to growing rice, a tropical plant. To carry the water from the forest to the rice, they dug irrigation channels. To this day, a water manager, who is typically a woman selected by the villagers and paid in rice, watches over the water flow, dredging and repairing the channels and making sure the rice receives its water.

But mountain regions are considered to be among the most fragile terrain on the planet, not exactly the place one would ordinarily choose to farm. The steep hillsides are prone to erosion and mudslides, and it's hard to maintain soil fertility when water is constantly flowing through your land, leaching the nutrients on its path down the slope. Farming at the bottom of the river valley would be a lot easier. That's one reason why people who live in mountainous areas around the world tend to be among the poorest on earth.[32] But mountains are good for security (you can see the enemy coming from afar), and historical circumstance and technology have allowed people to live here in Yuanyang, with their rice, for thousands of years. In fact, they were so successful at producing rice in this landscape that, centuries ago, the terraces covered more than eleven thousand hectares, with more than three thousand terraces stretching like a giant staircase up the steep hillsides. Towards the end of the fourteenth century, during the Ming dynasty, the terraces of Yuanyang became known as the Eastern Grain Barn.

People could transform this unlikely land into a rice-producing farm because they nurtured its biodiversity. They respected the variety of life, its ecosystem and that made it stronger.

By about nine o'clock the fog began to lift and the terraces came into view. It was early November, the beginning of winter, and for the most part, the rice had been harvested. The terraces were flooded with water, and from a distance, the ridges of their banks gave the impression of being ripples on the surface of an enormous pond cascading down the mountainside. The paddies, emptied of their rice for the winter and filled with water, had been stocked with fish, and there was lotus growing too, a flowering marine plant that is raised for its delicious roots and seeds. As long as I kept my back to the road and looked out over the undulating terraces, it appeared as if life here had continued uninterrupted by modern times. A water buffalo moved slowly through one paddy. A human figure bent over in the distance. Ducks frolicked.

This system of terraced farming hasn't changed much over the centuries. Because Yuanyang County is so remote, there has been little influence from the outside world, and most people still depend on traditional rice farming for survival. Unfortunately, the villagers are among the poorest people in China, and they have a hard time growing enough rice to feed themselves—in large part because there is a finite amount of arable land and the population has grown dramatically since the 1950s. The industry that has enriched other parts of the country never arrived in Yuanyang. The local road system was built only in the 1990s, and the terrain isn't suited to factories or industrial farms. When government officials visited in 2000 with plans to boost the local economy by increasing tourism, they found people living in a village only a few kilometres from Yuanyang who existed so apart from the rest of the world that they had never even seen a camera. Since then, a rise in tourism has boosted the economy so that today in this same village there is good cellphone reception, and in Yuanyang there are solar water heaters on the rooftops of many buildings. Nevertheless, globalization and its partner, industrialization, still haven't had too much of an influence. The majority of the women in the area continue to wear the traditional dress of the two main ethnic groups, the Hani and the Yi, and it's difficult to find a Coca-Cola to buy. Farmers rely on water buffalo and their own muscle, not on tractors and fossil fuels, and to this day, the water that irrigates the rice terraces is revered. People believe that special gods see over the water sources in the forests, which they still consider to be sacred groves where people aren't allowed to go and where hunting and logging are forbidden. The communities appear to be integrated into the natural cycles of life and, just as their ancestors would have been, protectors of biodiversity.

Not all is as it appears, though. From the vantage point of an outsider looking in, it's easy to be fooled by the bucolic setting, the water buffalo, and the lack of Western soft drinks. But just like every other place on the planet in the twenty-first century, this remote area has changed. People in these parts may rely on subsistence farming and grow their own rice to eat, but life certainly hasn't been static. To begin, Yuanyang is in China, and these days, China is synonymous with change. The country is experiencing what is said to be the world's largest ever human migration, as millions of people leave farming areas like this for the factories of Guangdong. From these rural hills, young men and women are heading to the industrial centres to find jobs sewing ski jackets and shoes or putting the parts in refrigerators and televisions, and when they go, not only do they leave behind their dependence on the land, but they leave their families and their traditions too. In one rice-growing village a short drive from Yuanyang, a twenty-two-year-old man wearing jeans, running shoes, and a red T-shirt told me that he had recently returned from Guangdong, several days' journey away from his home, where he worked in a factory. He said that's where he picked up the taste for a different kind of rice—the hybrid rice varieties bred by the government and the big Chinese seed companies. He now prefers it to the traditional kinds his parents grow. "It doesn't taste as good," he said to me of his parents' old rice lines. He said he prefers the newer, sweeter ones for sale in town.

In this man's village, Qinkuo, a tiny place that local officials have fingered for tourism development, he explained that most of the farmers still grow traditional varieties of rice. He pointed to a large sack of seed in a room on the ground level of their small house that overlooked the terraces and explained that the rice inside the canvas was of the same stock that had been passed down from generation to generation—the relatives of the rice plants his grandparents and his great-grandparents would have harvested. However, this is starting to change too. Whereas people in his village have, over the last few years alone, acquired televisions, cellphones, and even cars since the influx of tourist dollars and money from factory workers like him, these are merely superficial changes compared with the profound transformation in the rice terraces.

The shift that has the potential to substantially alter life here is the decision by some of the villagers to forgo their old seed lines and start to grow new high-yielding hybrid rice varieties. This decision, to plant one kind of seed over another, is of fundamental importance because it triggers a move away from the traditional farming systems. These new seeds aren't like the old ones. They require fossil-fuel-derived fertilizers to feed their growth, not the manure that the farmers are accustomed to using. To fight the insects that like to feast on the high-yielding hybrid rice, the farmers are instructed to use chemical pesticides. And when they use chemicals in their rice paddies, they can no longer raise the fish they typically farm in the terrace water, or keep ducks or harvest the wild foods that grow there. In short, everything about the way they farm must change. The biodiversity that was the foundation of this way of farming rice is traded for technology. At a farm store on the main street of old Yuanyang there are signs of this transformation. There, alongside a row of portable rice threshers, were at least a dozen pesticide-application packs you strap to your back and operate with a handheld pump to spray the terraces. There were also seeds for sale that had been brought in from elsewhere.

Biodiversity is all the life on planet Earth. It is the millions and millions of species that have evolved over four billion years. Technically speaking, there are three categories of biodiversity. There is the diversity in our DNA—all the genes that make up each human, each plant, each insect, each micro-organism. There is the biodiversity of different species, and there is ecosystem diversity—that is, all the variety of habitats such as rainforests and deserts and grasslands and oceans. What makes life on earth so beautiful, so alive, is the way all this life—this biodiversity—interacts. The connections and dependencies and interactions between species and habitats are so complicated that we don't fully understand how they work. What we do know, however, is that each species, each bit of life, has its own role to play. If you remove one bit of life from this mosaic, the effect can ripple out. That is what is happening around the world today. We humans, with all our industry and farming and living, are destroying this biodiversity. We are causing species to go extinct at a rate of one thousand to ten thousand times faster than the natural extinction rate. This too is happening in the rice fields of Yuanyang.

Farmers in the area used to grow hundreds of varieties of rice, varieties adapted to the different microclimates and local pests and diseases. That number has now dropped to the double digits. Whereas the rice terraces of Yuanyang were once richly biodiverse, this variety of life is dwindling quickly. It's a silent change, almost invisible to the outsider, but one so serious that it puts at risk the kind of farming system we need to have if we are to feed the world in 2050.

What's happening to the rice seeds of Yuanyang is not unique. This is just one of the many chapters in the larger story of what is happening to the seeds that grow our crops. Whereas farmers for the last ten thousand years have relied on food plants with a rich array of genetic material that produce a dizzying range of qualities, over the past century we've significantly narrowed this gene pool. We've gone from growing countless varieties of food crops, each thriving under different weather conditions and each offering different flavours, to producing only a handful of varieties selected for their uniform growth in the monocultures of industrial agriculture. An example of what this historic variety in food genes can look like is the dozens of kinds of apples that thrived in North America just fifty years ago. We used to have early eaters that ripened in August and keepers that lasted until spring. There were bakers for pies and white-fleshed fruits that liked the summers hot and yellow-skinned transparents that fruited very early in the year. In the mountain ranges of Kazakhstan, where it is believed the apple originated, there is an even more astounding range of domesticated apples as well as wild ones that offer a huge amount of genetic variation. Similarly, there are many varieties of garlic, wheat, corn, rice, onions, chilies, carrots—anything at all that we eat.

When we stop growing these varieties and we don't save our seeds, the genetic material they contain is lost forever. The diversity of our food crops plummets. According to the Food and Agriculture Organization, 75 percent of the world's crop diversity was lost in the last century. In communities around the world, replacing locally adapted varieties such as the old rice lines in Yuanyang with newer high-yielding varieties, some of them designed after the green revolution for industrial-scale agriculture, winnowed down, by more than half, the number of varieties of the world's three most important crops, wheat, corn, and rice. In India alone there were thirty thousand wild varieties of rice in the 1950s; by 2015 that number is expected to be fifty.[33] These are profound changes that affect our ability as a species to produce food to eat. Without seeds we can't grow anything. Without genetic diversity in our seeds, our crops lose the ability to adapt to the changing environment. And when these plants are unable to adapt, and are ravaged by disease or pests they can't fight, or aren't strong enough to survive droughts or floods, we don't eat. The health of our seeds—their biodiversity—is intrinsically linked to human survival.

Today, our food plants are at risk because of their uniform genetic makeup. When we lose the genes that express all sorts of helpful traits, the plants become genetically vulnerable. Not only do we lose individual genes but we also lose the genetic combinations that result when genes are expressed together. We've already seen the damage this can cause. In the 1960s, the commercial banana trade in Central America was almost wiped out because all the plantation banana trees were of the same gene stock and were defenceless against Panama disease, a fungus that rotted the stems and killed the plants. The industry was saved by an Asian variety, the Cavendish, that was believed to be resistant to the disease. (It was resistant only to that one strain. Today the banana industry is on the edge once more because the bananas grown on plantations around the world are once again genetically identical, putting them all at risk of a different strain of the same fungus that is now making its way around the planet.)

And we are not only losing variety in the DNA of our food crops. The wild relatives of our crops are also dwindling. That means the sources for new and as-yet-undiscovered beneficial traits are going extinct, usually because of human encroachment on the wild environment where these plants thrive. Maybe the genes that could save the banana from the aggressive fungus can be found in a tiny population of wild fruit trees in Indonesia—right where developers may be about to build a new hotel.

At the centre of this story of seeds is rice. Rice is life. People say it is the most consumed grain on the planet, with more than half of humanity relying on it every day. It's been important to society for millennia. About ten thousand years ago, when humans made the transition from being hunter-gatherers to farmers, rice was one of the first crops to be cultivated. Archaeologists believe domestication of the wild tropical plant took place as far back as eleven thousand years ago, in what is now China. Since then, generation after generation of farmers have collected the seeds from their best plants, choosing for traits such as taste, culinary qualities like stickiness, and colour, as well as resistance to environmental pressures, all to develop the hundreds of thousands of heritage rice breeds known to humans.

This century, however, rice production is faltering. Despite the boost in yields offered by the green revolution, today global rice production is on the decline. Disease and pests have caught up with the green revolution's high-yielding hybrid seeds. Severe depletion of natural resources in the rice paddies, such as soil salinization and environmental pollution, has further threatened crops. At the same time, the number of people who rely on the grain is growing. In Asia today, one hectare of rice feeds twenty-seven people. By 2050, this same hectare will have to support forty-three.

How can we feed a growing world population if rice is slipping? The answer may be found in the seeds themselves.

Cary Fowler is the former executive director of the Global Crop Diversity Trust, an independent organization based in Rome that is collecting money from governments and the corporate sector to create an endowment fund to preserve as much of the genetic heritage of the world's food plants as possible, in perpetuity—before they become extinct. What makes preserving this genetic material all the more critical, Fowler says, are the changing weather patterns around the world that result from climate change. "Everything is at stake," he said. "Climate change has presented humanity with a problem more severe than anything we've ever faced as a species. It's the central issue of our times. If I look at the history of agriculture, in the entire history dating back to the Neolithic period, I don't see the combination of challenges facing agriculture today. And we are woefully unprepared to face it now," he said. "If crops don't adapt to climate change, agriculture won't adapt. And if agriculture doesn't adapt to climate change, nor will we."

If the goal is to create a sustainable food system by the year 2050, then by the 2030s we must transform our relationship to seeds. Once we've met the first milestone of transforming the agricultural systems of the world to sustainable farming methods and have created the food systems that support them, the farmers behind these operations will need a supply of seeds that can meet the challenges of the decade. They will need a supply of seeds that is genetically diverse, is easily accessible, and will allow them to adapt to a changing climate as well as to produce enough food to nourish us all. They will need the support of biodiversity. For this reason, the second step in building a new food system is to preserve the genetic heritage of our food, thereby protecting biodiversity and ensuring that all farmers have access to the seeds we need to feed the world in 2050.

CHAPTER EIGHT

The Genes in Our Seeds: The Big Business of Food Security

The journey to Yuanyang was an adventure. I spent the minivan ride clutching the strap above the window, staring at what could be seen of the road ahead as it curved around the next ripple in the mountain. I should have been admiring the view of the rice terraces we were passing, but if I looked out the side window, I could only focus on the tiny space that separated the wheels of the van from the precipice beside what sometimes thinned to a dirt track. I didn't enjoy the ride. My travelling companion and guide, Rose, however, was calm. She was used to the rugged terrain and laughed as we bounced so hard our heads hit the van's roof.

I had met Rose in Kunming, the capital city of Yunnan. She was introduced to me by my friend Tony Fuller, a retired professor of rural planning and development at the University of Guelph who now lives in Beijing and had invited me on this trip. Rose is a tour guide who accompanies Western tourists around Yunnan, the province where she grew up. She is so knowledgeable about the area that she often sounds like a living guidebook, reciting statistics about population, sharing factoids about the natural world, and interpreting not only the language but the cultural landscape—such as informing us that we would stop at a Muslim restaurant for lunch because in Yunnan, restaurants owned by those who practise Islam are thought to be cleaner. So in we went to a dining room off a large courtyard and were invited to choose the ingredients we wanted cooked for us from a display of mushrooms, fresh vegetables, and tofu stuffed with meat and an assortment of animal parts. Within fifteen minutes, we were served dishes of spicy stir-fried soybeans, tofu, and green beans by a young Chinese woman wearing the hijab.

In some ways, Rose and I are quite similar. At thirty, she was only a few years younger than I was, and we are about the same height (short). We are both mothers to young daughters. She would have liked a second but couldn't have one because of her country's one-child policy. Thanks to globalization, we wore similar outfits: knee-high boots, wool coat, hair pulled back in a ponytail. Yet while I'm from downtown Toronto, Rose came of age in the rice terraces of Yunnan. Rose, it turned out, was the ideal guide for this rice tour. Not only did she know all about Yuanyang but she was also knowledgeable about rice cultivation. She knew from first-hand experience all the hard work it takes to steward the seed from germination all the way to harvest. I eat rice, casually, without thinking much about how each grain is grown, whereas Rose had watched the seeds sprout, transplanted the seedlings herself, and then later helped her family bring in the crop.

As we travelled together, I learned that Rose's story told the bigger tale of what's happened with seeds and agriculture in small peasant families like hers as more and more people make the transition from traditional farming methods and adopt the tools of industrial food production. What will happen in the next few decades on farms like Rose's family's paddies—like Chandrakalabai's in India, like Bryan's in Ontario—will determine the future of our food systems and their impact on the environment and on climate change.

Rose grew up in the Yunnanese rice terraces, a short distance from the Burmese border. Like the majority of Chinese in the 1980s, her parents were subsistence farmers, struggling to survive. They lived in a remote village, a seven-kilometre walk from the nearest road past rice terraces, tea fields, and water buffalo. The village was so small there wasn't even an elementary school, and at six years old, Rose was sent to the government boarding school in the nearest town. She would see her mother on Tuesdays, when she came to sell her vegetables and dried firewood at the market to pay for Rose's tuition. "By the time my mommy visited me each week, she ran to me with tears," Rose remembered. "The most happy day was market day. My mommy finished selling vegetables and came to see me. That was the most happy day." Then on Saturday, Rose and the other children from her village would walk those seven kilometres all by themselves back to their homes, to spend one and a half days with their families before returning once again for the school week.

As per the government's socialist policies of the day, land in Rose's corner of Yunnan was shared among those who lived there. The families worked together primarily to grow rice. During this period, with communal work brigades cultivating the land, the country's agricultural systems didn't thrive. Not only were yields low, but the government took away a proportion of rice to meet production quotas, which left little of the crop for those who had produced it. That meant hardship and starvation for the first years of Rose's life, and she remembers being hungry. "We had less to eat because we didn't have our own land," she said. The area where they lived was predominantly populated by the region's ethnic minorities, and even though Rose's family didn't belong to any of these groups, the government's one-child policy didn't apply because ethnic minorities in Yunnan remain exempt from this population-control measure. Rose's parents were permitted to have another daughter and then a son. So when there wasn't enough rice to feed the family, Rose's mother mixed the girls' share with the wheat or corn, but fed her son pure rice.

At some point during her childhood—Rose can't remember exactly when—the government land reforms that began in 1979 were instituted in her area, de-collectivizing agriculture. Her family was allocated their own plots. Rose might not remember exactly when this happened where she lived—the reorganization of land took place across the country over a number of years—but she vividly remembers its effect on her family: the females began to eat rice. "It changed a lot," she said. "Since we have land, we all eat pure rice every day. You have land, you can grow whatever you want. You have more to eat. If you are not lazy," she added. "Both my parents worked very hard." After the land reforms, she said, "we had vegetables, sticky rice." Her mother would even make a traditional sweet dish from sticky rice that she fermented with sugar in a clay crock. "It was wonderful."

Rose's family were not only hard workers, they were excellent farmers too. They were industrious and smart, and their farm quickly prospered. In the dry season, they planted wheat and grew the region's specialty, pu'er tea, picking the leaves and selling them to a local factory. They also sold at the market the wheat and vegetables they didn't eat themselves, and they grew several varieties of rice. There was traditional rice and a sticky rice that clumped together when it was cooked and was used to make the sweet sticky rice dish at holiday times such as Spring Festival. They also grew a short-grain rice for everyday consumption and that Rose's mother fermented to make a cherished alcohol. When the paddies were flooded, they also used them as ponds to raise fish that they would catch at the end of the season and then ferment and turn into a paste to flavour their foods. Not everyone was as lucky as they were, said Rose. There was at least one family in the village that wasn't able to prosper like her parents and who struggled to grow enough to eat. But for her family, the often back-breaking work they did farming provided a good living.

Rose remembered her years in the village with the warmth and fondness of someone who has left it behind for a new life in the city. One night we were driving on the highway through Yunnan, not long after the sun had set. From the window of the van we could see the silhouettes of the farms we were passing against the fading sky. We were talking about her time in the village and I asked Rose what image came to her when she reminisced about her hometown. She closed her eyes. "In my heart, it is harvest season. The rice plants are drooping very low with the seed and they are a golden colour. That is the best picture in my eyes."

Farmers like Rose's parents have been practising small-scale peasant agriculture, saving seeds to sow and propagate food crops, for what historians believe has been about ten thousand years, ever since our ancestors made the transition from hunter-gatherer to farmer. It has been only during the last hundred years or so that we have broken with this history of seeds and have formalized our seed breeding, beginning with changes made more than a century ago. And this break has come at a great environmental and social cost that puts our global food security in jeopardy as we move towards 2050. The damage we've done to our biodiversity and the way we reproduce, buy, and sell seeds today stands in the way of building sustainable food systems. To better understand what's happened, we need to look back at how seed breeding has changed over the millennia. Once we look at where we've come from, it will be easier to see where we are headed. We can more clearly understand the role seeds and biodiversity play in this narrative of farming.

No one knows for certain where, or how, humans first began to farm, but it is believed that agriculture developed independently in several parts of the world, at about the same time that the globe began warming after the last ice age. It's generally thought that farming began simultaneously in Ethiopia, Mesoamerica, and central Peru, as well as in the Middle East, Southeast Asia, China, and New Guinea. The transition from human groups who survived by harvesting edible wild foods and hunting game to farming communities where people produced what they ate took place over a period of about five thousand years, which is a flash in evolutionary terms, particularly because *Homo sapiens* had been surviving on the planet without farms for more than two hundred thousand years before that. But for some reason, people in what is modern-day China started to grow rice. The ancient Peruvians planted their potatoes, and Ethiopians their wheat. The first farms took shape, and agriculture was born.

At the base of this profound shift from collecting food to cultivating it was the process of plant domestication. People probably saw characteristics in wild plants that they liked—maybe they tasted really good, or filled the stomach well—and tried to replicate them by collecting and sowing their seeds, transplanting their suckers or seedlings, or harvesting their tubers and burying them. To become farmers, though, they had to give up their nomadic hunter-gatherer lifestyle and the security it provided. That was, historians say, a better life than farming because starvation was rare, hunter-gatherers needed to work less than a farmer does, and the food they ate was more nutritious. So if they were to tie themselves to one spot and not starve, humans had to transform these wild plant species into reliable food-producing varieties and adapt them to the growing conditions the early farmers provided. (Of course, this process took place with the animals we would herd and the birds we would raise in addition to the plants we would cultivate, but for the purposes of this story of seeds, I'm going to focus on what botanists call the flowering plants, a scientific category that includes tubers like potatoes, grasses such as rice and wheat, and other plants that propagate via suckers, such as the banana.) In this way, humans began what would turn into a millennial relationship with all this biodiversity of these other life forms.

When we humans started to produce our own food, we became agents of evolution, dramatically altering plants by choosing to save one seed over another. We changed entire plant populations. A good example of this is the case of corn, or maize. The ancestor of today's corn is a weedy grass called teosinte, which is indigenous to Mexico and Central America. Corn was likely first domesticated between seven and ten thousand years ago by the indigenous people who lived in the part of Mesoamerica that is now Mexico; from there, the crop travelled north and south. (There has been some debate over the years in academic circles about the origins of corn, but DNA studies seem to point to nine thousand years ago.) If you compare the DNA of teosinte and today's corn, there are few differences. However, the wild teosinte looks nothing like corn. Teosinte is a scrubby, bushy grass, with nothing resembling corn's uniform leaves and long, sleek stalks that stand in straight lines in farmers' fields. The most significant difference between the two, though, is the seed heads—the flowering part of the plant, its reproductive organ, which grows into the corn we eat. Whereas maize produces giant seed heads, corncobs heavy with kernels, teosinte's are merely a fraction of the size. Through domestication and careful seed selection, humans transformed the teosinte into a plant that produced more food.

Probably the most important domesticated crops were the grasses, grains such as rice and millet. Over millennia, farmers have developed all sorts of varieties of these edible plants, varieties that exhibited incredible biodiversity at a genetic level. This was a security measure. They didn't know it at the time, but genetic diversity is a kind of insurance policy. It offers the raw material for the plants to adapt to a changing environment. The more diverse a gene pool, the more likely it is that a population of plants will have a gene that provides resistance to a range of stresses. If growing conditions were dry one year, then seeds from the plants that survived the drought would end up being the ones saved for the next year. The following summer, the plants that thrived despite the onslaught of, say, a fungus or predatory insect would be the ones collected. The year after that, the genes of plants that reproduced despite heavy rains and flooding would be preserved in the saved seeds. In this way, farmers created crops that expressed a variety of traits to ensure that people would have something to eat no matter that year's growing conditions. They didn't nurture only one kind of rice but rather created thousands of varieties, each suited to the particular geographic area where it was grown. These locally adapted varieties are called landraces. We humans were so successful at breeding these that, until recently, there used to be more biodiversity within agricultural plant communities than there was in their wild relatives.

One reason for this is that landraces are in a constant state of genetic flux because their genes are always interacting with their environment. For example, in farming communities near wilderness areas where the wild ancestors of domesticated plants grow, the distant cousins crossbreed. This introduces new genes into the domesticated stock in what's called introgression. Which is a good thing, because just like plants in the wild, domesticated crops are in an unending evolutionary struggle with pests and disease and are constantly adapting to changing weather patterns. A diverse gene pool helps them out. Plants need an evolutionary toolkit to outwit their opponents, and genetic diversity is the key to their survival.

When we bred seeds for foods we liked, we took them with us when we travelled. Wherever we migrated, over mountain ranges and across oceans, we packed our seeds. When some of the first farmers set out from the comfort of the Middle East's Fertile Crescent to explore new territory or to trade, we know they carried their seeds with them because wheat was being grown in other regions a few thousand years later. The Vikings, known for their conquests of territory as distant and uninhabited as Iceland and Greenland, would have taken with them the seeds for their favourite crops to plant as well as livestock.

Cauliflower was first bred in what is Lebanon today, but in the late Middle Ages, Italian traders took the vegetable across the Mediterranean to Europe. The original carrot, descended from a wild relative in Afghanistan, first appeared in tenth-century Turkey. Within two hundred years, these edible roots were being eaten in Spain and later in Asia, making it to China by the fourteenth century and into Japan by the seventeenth century. The first carrots would have been yellow or purple. They weren't bred to be the orange spears we eat raw with dip until farmers in the Netherlands selected for that colour at the beginning of the seventeenth century. The most significant movement of food-crop DNA happened when Europeans landed in the Americas in the fifteenth century, prompting the world's biggest ever exchange of plant DNA. Crops from the New World such as tomatoes and potatoes were imported to the Old, where they were quickly adopted, and colonists migrating to the Americas took with them their favourite plants to grow, such as wheat.

Farmers might not have understood the science behind what they were doing, but their success wasn't haphazard. As horticulturalist Noel Kingsbury describes in his book *Hybrid,* which covers the history of seed breeding, various societies throughout history have embarked on seed-breeding programs to improve farming systems and to support rising populations. In China, during the Song dynasty (960–1279), the government launched a campaign to improve rice yields to feed a growing populace. It distributed early-maturing rice varieties, developed by farmers in Vietnam, and created official postings for master farmers, who taught other Chinese farmers about irrigation and fertilization. These state efforts led to rice surpluses. Then in the late 1800s, with one discovery, seed breeding took a turn in a new direction. An Austrian monk named Gregor Mendel laid the groundwork for modern genetics and the science-based plant breeding that launched the critical loss of biodiversity that exists today and that is one of the most serious hurdles we face for the future of food.

These days, we learn in high-school biology about Mendel's experiments with pea plants that proved the laws of inheritance, but less than 150 years ago, humans knew that traits were passed down from one generation to the next, but they didn't understand how physical characteristics like colour or leaf shape, or even resistance to disease, could be inherited. Darwin and other thinkers such as Alfred Russell Wallace were only just articulating the concept of evolution in the late 1800s (and were greeted by much scepticism from the scientific community). What Mendel showed by breeding and observing thousands of pea plants was that a plant's physical characteristics, called phenotypes, are passed down from one generation to the next in a predictable fashion. He demonstrated that the two parents' physical makeup determines their progeny's makeup. He also observed that the offspring can have a combination of characteristics that exists in neither parent.

Mendel worked in isolation, and many of his notes were destroyed after his death, so it wasn't until several decades later that his ideas were discovered, embraced by the scientists of the day, and then applied to plant breeding. The effect was profound. Plant breeding was transformed into a science. Whereas until then, seeds had been propagated only by farmers in their fields, from that point on, much plant breeding would be carried out by professionals in laboratories and on research stations. In the words of Kingsbury, plant breeding was "transformed from craft to industry." Mendelian genetics inspired scientists to dream of creating seeds that would produce the same results year after year. Unlike the genetic variety in a landrace that causes plants to change slightly every season, the new seeds they wished to breed would be pure lines with predictable results. Using today's vocabulary, they wanted to create high-yielding hybrid seeds.

To produce hybrid seed, you need to first create what's called a pure line, a plant that produces exact clones of itself when it is inbred. You do this by inbreeding a crop for several generations until you have a group of plants that are genetically identical. Next, scientists experiment by crossing two pure lines together. The theory goes that the offspring of these two pure lines will exhibit what's called hybrid vigour. That is, they will be all around better than their parents and have higher yields. But this benefit lasts only one generation. This means that farmers who plant this first generation of seeds can't save the seeds of their offspring and hope to get the same kind of yields the next year. The grandchildren of those two pure lines will perform poorly and offer uneven yields. This has been described as a biological lock. The farmer who plants hybrid seeds has no choice but to return to the seed company and purchase new seeds every year, something the seed companies have been severely criticized for.

Still, these types of seeds have been extremely popular. Farmers around the world quickly adopted the new hybrid seeds, beginning with hybrid corn in the 1930s. The majority of farmers in North America and Europe now plant hybrid seeds, and the large, industrial monocrops of the developing world also depend on them. Yet it is their very popularity that has caused the loss of biodiversity, because farmers abandoned the old seed lines to plant the new hybrids.

Back in the 1920s and '30s, the loss of genetic material was invisible to the majority of scientists. But one man did notice. Nikolai Vavilov was a successful Russian botanist whose legacy for humanity was appreciated only years after his death. Vavilov travelled the continents in search of what he named the "centres of origin," the places where agriculture was born. His theory was that the best place to look for crops with immunity to disease and pests would be the place where that crop had originally evolved. His expeditions took him across continents, from China through the Middle East, into North Africa and south to Ethiopia. He went to Italy, Greece, and Spain, and later in his life he visited the Americas, traversing a great expanse from the Brazilian rainforest all the way north to the American Southwest. And everywhere he went, he collected seeds of landraces and interviewed farmers about their techniques. Back in Russia, the seeds he collected were gathered in a scientific collection that is still one of the largest collections of plant genetic material. He hoped he would be able to tap into their genetic diversity to improve agriculture and breed better seeds to help feed his country. According to ethnobiologist and author Gary Nabhan, who wrote a book tracing the botanist's journeys, Vavilov was the first person to articulate the connection between the biodiversity of our food and our ability to feed ourselves. The crux of his theory was that agricultural biodiversity is the foundation for human food security—a realization that is even more profound today.

It was in the early 1960s that an increasing number of scientists began to notice the kinds of changes Vavilov had first spotted. Botanists, geneticists, and other academics working in places such as Iran, Turkey, and India witnessed a dramatic drop in genetic diversity in agricultural crops. At a 1967 conference hosted by the Food and Agriculture Organization, the scientific community came up with a word to describe the global phenomenon. They coined the term "genetic erosion" to mean the irreversible shrinking of the gene pool out of which new varieties should emerge. The problem was widespread. The same loss of biodiversity was happening in the richest countries, where industrial farming had already taken off, as well as in the developing countries that were importing new seed lines. In Turkey, almost all of the country's landrace wheat crop had been replaced by modern varieties. In Southeast Asia, the mixed farms that had existed for millennia were being replaced by monocrop plantations of oil palm and rubber trees that obliterated biodiversity.

Vavilov had offered the world a way to save these precious resources by collecting and preserving them. He had intended to work with his global seed collection back in the USSR, and he and his colleagues had catalogued his finds in the seed bank that still exists today, at the N. I. Vavilov All-Russian Scientific Research Institute of Plant Industry. But Vavilov never had the chance to finish his work. He was imprisoned by Stalin after a politicized scientific feud and died of starvation in jail in 1943, at the age of fifty-six. However, his legacy does live on in seed banks around the world inspired by his incredible collection.

Cary Fowler, the former head of the Global Crop Diversity Trust, the international organization that is raising funds to preserve agricultural biodiversity in perpetuity, has worked with the people safeguarding the biodiversity of our seeds today. "What we are doing is trying to preserve options," he said. "We don't have to predict in what way the environment will change, or the way the food system will change. We don't even need to predict who will produce the seed. All we have to do is predict that there will be change. Conserving the diversity is a no-brainer. This is the common heritage of mankind. What is going to be necessary are varieties that are adapted to the new climates that are coming. Where does that adaptation come from? It comes out of the diversity of crops that already exists."

There are two ways to preserve the DNA of our seeds. The first is called in situ conservation, whereby farmers grow varieties in the fields and wild species are protected in their natural habitat so that the genes can constantly evolve and interact with their environment. Then there are seed banks, also known as ex situ conservation. A seed bank is like a living archive, a place where samples of genetic material are preserved. Some seed banks are big, some are small. Some are run by villagers with few resources and whose collections are planted out in the fields every season. Others are large government-funded enterprises with cryogenic freezers and all sorts of modern equipment designed to preserve DNA forever. Some of the world's institutional seed banks are run by the Consultative Group on International Agricultural Research, an umbrella organization based in Rome that operates out of fifteen centres in different countries to promote agricultural research that reduces hunger and supports ecosystem resilience. Its gene banks preserve more than 650,000 samples in the public domain.

But not all food plants grow from seed, which complicates the task of preserving their genes. Potatoes, for example, grow from tubers—each of the potato's "eyes" becomes a shoot for a new plant. And bananas regenerate through suckers, baby shoots the plant sends out underground. To keep a banana genotype, scientists must preserve that sucker's growth tip. The inexpensive way to do this is to keep it in a test tube, but the specimen needs attention every year as it continues to grow. If you want to keep the banana genotype safe for a long time, another option would be to remove a microscopic cube of growing material and stick it in liquid nitrogen, where it will keep indefinitely. This is also the hard way. Charlotte Lusty, a scientist at the Global Crop Diversity Trust, has performed this type of tissue culture. "It felt like trying to peel a leek with knitting needles," she said.

Preserving the genes of our food in an institutional setting has been criticized. There are some who believe that gene banks provide corporations that develop and sell seeds, such as Monsanto, free and unencumbered access to the genetic material that indigenous people around the world have developed over millennia without compensation. They believe that the wealthy North has benefited from the incredible genetic diversity of the South without offering anything in return except policies that promote industrial agriculture and lead to dependence on patented seed lines and costly technology. (Even Vavilov has been accused of wandering about and taking away the natural heritage of indigenous people.)

Seed banks also cost a lot of money. Most, if not all, are plagued with funding issues. The FAO's *Second Report on the State of the World's Plant Genetic Resources for Food and Agriculture,* a document that examined the global state of seeds and biodiversity, described a sector in a chronic state of underfunding. Seed banks are hardly a government priority. Also, they can be victim to natural disaster, malfunction, power shortages, and even politics. During the revolution in Egypt in early 2011, seed banks in the Sinai were pillaged and years of work destroyed. They are also vulnerable to the whims of politics and money. The Global Crop Diversity Trust was trying to prevent the Pavlovsk experiment station near St. Petersburg in Russia from being turned into luxury homes. At stake were more than three thousand varieties of fruit and berry plants preserved in its gardens, as well as other accessions. So far, the government has halted the project, but the Trust is awaiting an official decree to ensure no one ever again tries to destroy the genetic resources held there. Because of the many inherent weaknesses in the gene bank system around the world, the Global Crop Diversity Trust has created a way to preserve all this biodiversity in one place: the Svalbard Global Seed Vault, on a remote Arctic island in the Norwegian archipelago of Svalbard.

There is something chilling about the idea of Svalbard. Perhaps it's the gene bank's purpose, which is to "provide insurance against both incremental and catastrophic loss of crop diversity" and to offer "'fail-safe' protection for one of the most important natural resources on earth." It is the world's only backup when it comes to saving our food. Or maybe it's the location that is discomfiting, way up at the top of the world, almost a thousand kilometres north of mainland Norway, a place so purposefully distant from most of the world's civilization. The mountain site was chosen because it is substantially above sea level and rising water can't reach it. The area is also geologically stable, and a 1920 treaty prohibits military activity in the area. If someone wanted to break inside the gene bank, it would be tough. They'd have to first get past the polar bears that roam the area and then the four locked doors that each requires a different key and security code. Not surprisingly, it's been called the Doomsday Vault, a name that conjures up images of future conditions so apocalyptic that in order to produce our food, and to survive, the remaining humans will be forced to travel to the Far North to procure seeds for their crops.

Photographs of Svalbard show a simple yet haunting entrance, the roof of the concrete structure filled with prisms and steel that acts as a beacon, lit by fibre optic cables in the winter and reflecting the polar light in summer. Behind this is a tunnel leading to three vaults carved into the permafrost and surrounded by thick mountain rock so that in the case of a power failure, the seeds will remain frozen. The Norwegian government paid the $9 million it cost to build the gene bank in an act of generosity towards the world, and the facilities are the property of the government of Norway. Operational costs are covered by Norway and the Global Crop Diversity Trust. Anyone is allowed to send seeds to be stored there—the service is free—and the seeds remain the property of those who send them. Teams of people around the world are working against time to send genetic information to Svalbard. Mexico has shipped the most accessions, followed by the United States and then China. Canada has sent some too, as well as countries such as Indonesia and Ethiopia that have fewer financial resources but substantial genetic ones. Peasant organizations have sent some of their own agricultural heritage.

Unfortunately, preserving agricultural biodiversity alone isn't enough to provide us with the seeds we will need to feed the world in 2050. That's because the way seeds are developed, bought, and sold needs to be overhauled. Seeds today are big business. Over the last forty years, large transnational firms have taken over the global seed industry from small family-owned companies. The market they've captured is worth more than $36 billion a year. Hybrids must be purchased year after year to achieve their advertised results, making them a profitable commodity. Up until the 1980s, these seeds were produced on the whole by small, often family-owned businesses, supported by research done by government extension programs. Then the large chemical and pharmaceutical firms began to buy up these small companies. Ownership in the seed industry consolidated through mergers and more acquisitions so quickly that by 2009, only a handful of companies dominated the global seed industry, distorting the market and limiting the choice in seeds: the US-based Monsanto (the largest seed company in the world), DuPont, and Dow; the Swiss-based Syngenta; and the German Bayer and BASF.

According to the FAO's *Second Report on the State of the World's Plant Genetic Resources for Food and Agriculture,* released in 2010, governments around the world have largely withdrawn from seed production and entrusted the field to the private sector. As a result, only seeds for the most profitable crops such as grains and hybrid vegetables designed for "market-oriented agriculture" are being propagated. According to the report, crops with less market opportunity, such as self-pollinated ones, are no longer being produced on a large scale. While many of the world's farmers still save and replant their own seeds year after year, trading them with their neighbours to maintain vigour, this practice is on the decline, particularly in industrialized countries.[34] In the 1960s in the United States, 63 percent of farmers saved their own soybean seed stock, whereas only 10 percent did so in 2008.

Today's seed industry limits sustainable agriculture in three ways. Firstly, as the market consolidates and the smaller seed companies disappear, entire seed lines are lost. Between 1984 and 1987, almost a quarter of the mail-order seed companies in the United States and Canada went out of business or were bought by bigger players. Many of the open-pollinated heritage varieties that were propagated by these small companies were discontinued. For the global companies, heritage varieties are nowhere near as profitable as hybrids because people buy them once and save them from year to year. It's also cheaper for a global seed company to breed one genetically uniform variety to service all markets rather than to maintain small seed lines adapted to a variety of regions. In 2000, the world's largest vegetable seed company at the time, Seminis (now owned by Monsanto), discontinued two thousand varieties of fruits and vegetables as part of a "global restructuring and optimization plan." The Seed Savers Exchange, based in Iowa, has tracked the disappearance of non-hybrid vegetable seeds in Canada and the United States with its *Garden Seed Inventory.* It has found that 57 percent of the roughly five thousand non-hybrid vegetables that were available by mail order in 1984 were no longer available in 2004. What this meant for cauliflower, for example, was a drop from 152 varieties available through these catalogues in 1981 to a mere 30 varieties in 2004. In the introduction to the 2006 edition of the *Garden Seed Inventory,* they conclude that backyard gardeners who plant what's left of these non-hybrid varieties are "the stewards of this irreplaceable genetic wealth." Food security and biodiversity obviously are not as important as profit.[35]

Secondly, as seed markets consolidate, farmers lose choice. One cotton farmer I met in India told me he planted genetically modified cotton seeds even though he was an organic farmer (GMOs are banned by organic standards and certification programs) because there simply wasn't another kind of cotton seed for sale and the heritage varieties people planted before had disappeared. The seeds are also often only sold packaged together with other products, including pesticides, reducing farmers' ability to choose what type of agriculture they want to practise. According to an article about the consolidation of the global seed industry published in the journal *Sustainability,* farmers who purchased new hybrid barley seed were forced to also pay for fungicides and plant growth regulators.

Thirdly, according to a 2004 study by the US Department of Agriculture, as the seed market consolidates, competition in the industry lessens, and so does research and development. This leaves us with an innovation gap, because public institutions that previously developed new seed varieties have largely been cut from government budgets.

This situation must change to support a sustainable food system. We must recognize that the preservation of biodiversity is more important than company profits. But supporting biodiversity doesn't mean going back in time. Technology and innovation can still help us.

CHAPTER NINE

Lab Rice: A Better Seed for a Hotter Planet

Far away from the rice terraces of China, in the basement of the Department of Ecology and Evolutionary Biology at the University of Toronto, about a dozen rice plants in a growth chamber transformed the artificial sunrays from the bright lights overhead into plant stalk, leaves, and seed. Tammy Sage, an associate professor of plant biology, pulled open the door of the growth chamber, which looked like a giant fridge. Instead of cold air falling out, I was hit with heat as if I were getting off a plane in a tropical country; the artificial sunlight was so bright I winced. The long and slender green leaves of rice plants poked out of holes in the lids of a plastic box where they grew in a solution of water and nutrients. The grain had formed, and the stalks drooped slightly under their weight. They recalled Rose's description of the rice fields near her home at harvest season, the plants golden and heavy with the ripe grain.

As alien as these growth chambers appeared to me compared with the rice terraces—their white walls and mechanical hum replacing the mud banks and the trickle of running water—they simulate conditions in the field, allowing scientists to observe how the plants grow. A touch screen was used to control the environment in the chambers, regulating the temperature for this particular experiment—a steady 18 degrees Celsius at night and 26 degrees during the day; it also kept the daylight constant, unlike in a field, where the sun can tuck behind cloud cover. This precision control of light exposure allows scientists to observe what effect the number of daytime hours have on a trait they want to investigate, such as a plant's flowering time. Sunlight, and conversely the length of a night, determines when a plant will flower. Scientists want to control every single variable in the plant's environment so they can explain the changes that take place. "You want to control the variables to understand how they influence plant growth," explained Tammy Sage.

The experiments in Tammy's lab focused primarily on the effect of temperature on the rice plant's reproductive processes. In another growth chamber down the way, the air was hotter, set to a steady 32 degrees. The rice plants there were young and hadn't flowered yet, but some varieties were expected to do so and then go to seed, while others would not, and this would allow her lab to observe how temperature influences plant reproduction. "Once we understand this developmental reason for why this rice is tolerant to the heat, then we can find the gene that determines the heat tolerance," she told me.

Rice is a tropical plant, one that thrives in the warm, humid climate of places like Yuanyang. But it doesn't like it too hot. As the temperature rises, rice reproduction starts to falter. Knowing what gene allows rice to survive and produce grain under hot conditions is important for our future. As climate change pushes temperatures higher, the rice varieties we plant will need to be heat tolerant if we want to continue to grow rice successfully for food. "At 32 degrees and higher, nothing grows. The pollen is not available for the fertilization process," said Tammy. When it is that hot, the rice flower develops but the pollen, which carries the sperm, aborts. And if there is no pollen, there is no fertilization. Without fertilization, there can be no rice grains.

Scientists at the International Rice Research Institute have been watching the effects of climbing temperatures on rice since the 1980s. For every 1-degree increase in the minimum temperature, rice yields decline by 10 percent. Rice is already suffering in this new and shifting climate. The Intergovernmental Panel on Climate Change predicts increases of between 2 and 4 degrees Celsius in coming decades. "Plants are hitting sterility thresholds even more. We need to understand how temperature reduces yields."

The rice varieties of the green revolution are also approaching what is called the yield ceiling, the maximum amount of rice that a plant is physically capable of producing when grown under optimal conditions. "This has been creeping up over the last thirty years as we bred better varieties that do things like capture sunlight better," said Rowan Sage, Tammy's husband and partner in science. "But now we are hitting the point where photosynthesis has maxed out." Farmers have been trying to compensate for this plateau in their yields by adding more and more fertilizer, to the peril of their environment. One reason they haven't been helped by newer hybrids, explained Rowan, is that scientists haven't been able to breed the plants to allot any more of their resources to reproduction, to producing grain. At the same time, agricultural lands have been turned into strip malls and homes have been built where people once grew food, soil has been degraded by pesticide use, and the water that has irrigated rice fields is drying up in part because of an increase in human water usage but also because of a shift in rainfall patterns associated with climate change. As well, many farmers aren't even able to achieve what scientists call attainable yields—the yields that a particular soil type and farming method should produce—particularly with rice in Asia.

For us to feed billions of people under the pressures of climate change, we need seeds for the plants that will be able to flourish under new climatic conditions. Which raises the questions of what these seeds are and where they come from. Has evolution provided us with the seeds we need—if we only could preserve their diversity from extinction? Or do we need to deploy all our human ingenuity and technology to have scientists create the seeds that can feed us by 2050? We must preserve biodiversity and reverse the negative fallout of corporate consolidation in the seed industry, but this should not stop us from drawing on science to help solve the question of how to provide food for humanity in a sustainable way.

Some of the world's top plant scientists hypothesize that the type of seeds that can feed the world under climate change can be developed by them, in the laboratory. One group, of which Tammy and Rowan Sage are a part, believes that they can breed a rice plant to create more rice grains by re-engineering—or bioengineering—its leaves to make photosynthesis more efficient. They believe they can improve its ability to turn the sun's energy into carbon or biomass—which is our food. This means rebuilding the leaf: changing its structure, reconfiguring its inner workings. The project is called the C4 Rice Project and its tag line is "using the sun to end hunger." It brings together more than two dozen scientists in ten countries who are specialists in molecular plant physiology, biochemistry, plant breeding, and plant development. Their goal is to "supercharge photosynthesis" in a bid to boost food production and, they hope, improve the lives of the billions of people who rely on rice. They've received funding from the Bill & Melinda Gates Foundation as well as other granting organizations such as the Canadian International Development Agency and the European Union. The Sage lab at the University of Toronto is one part of the complex, worldwide network that is pooling resources for this project that is so big it will likely not be completed—if it ever is—in the lifetimes of those who launched the quest. In fact, this project is possibly one of the most difficult biological puzzles ever attempted, and it shows us that hope for the future might be found in a laboratory.

Tammy and Rowan Sage first met in high school in Nevada and have had parallel careers since grad school, though she is a plant cell biologist with a doctorate in plant physiology and he studied botany, his speciality being photosynthesis. Tammy is a beloved professor who has won awards for her teaching and who also runs the lab at the University of Toronto, while Rowan is top of his field and is world-renowned for his work in photosynthesis. They were conferring by his desk when I arrived for our first interview, and when we moved to Tammy's office to talk about the project, Rowan came along in stocking feet. There is a comfort level that comes with working with one's spouse.

The idea behind creating a higher-producing rice plant was born out of the belief that the best way to break through the yield ceiling is to transform the plant itself. What hubris! To re-engineer the leaf? The leaf that is one of evolution's most astounding creations, with its ability to turn sunlight into food that supports all us heterotrophs—we species that cannot produce our own food and so must eat other organisms for our nourishment. It sounds, well, crazy. But the idea is in fact inspired by nature itself. Evolution has already created a more efficient photosynthesis. About 90 percent of the plants on earth, including rice, use what's called the C3 photosynthetic pathway to transform the sun's energy into biomass, or plant matter. Most of the other plants have evolved beyond C3 to use a different anatomy, called the C4 photosynthetic pathway. It's a bit like the difference between a propeller plane and a jet plane, explained Rowan.

The first version works well—it takes off and lands smoothly—but the newer version is just that much more efficient at accomplishing the same tasks.

Cyanobacteria were performing photosynthesis billions of years ago. Then, sometime around 1.6 billion years ago, a host cell in an ancient heterotrophic organism engulfed a certain cyanobacterium. It survived inside this organism, and the two evolved to form a symbiotic relationship. The cyanobacteria lost their distinctiveness as organisms living within another organism and became part of their host; they became the organelle within a plant's cell that we call the chloroplast. These are the cellular compartments that cause a leaf to look green. Today, that's where plants make food from the sun.

All plants have evolved to have chloroplasts, regardless of whether they have the C3 or C4 photosynthetic anatomy. Although a fewer number of plants transitioned from C3 to C4 photosynthesis, the new and improved varieties nevertheless cover a quarter of the planet's surface, mostly in warm and tropical climates. Their advantage is that they are more efficient at converting light energy to carbon at high temperatures. Some of the world's most important crops, such as maize and sugar cane, are C4 plants, whereas rice and wheat are C3. That said, C4 are not implicitly better plants. In an ecosystem there is a role for everything. So, for example, you would find C3 plants only on the shaded forest floor or in a northerly woodland because the C4 anatomy doesn't perform well in those colder environments or where there is less sunlight.

C3 plants convert a smaller proportion of the sun's energy into plant material, or biomass, because they have to perform photorespiration. In these plants, photosynthesis creates a chemical by-product that is toxic to the plant in large amounts, and so the plant must get rid of it. Through photorespiration, the plant metabolizes this toxic compound, converting it into a carbon molecule that can be recycled using some of the energy it already captured from the sun. All this means that there is less energy left over after photosynthesis for the C3 plant to use to grow. "C4 plants have solved photorespiration," said Rowan. They've evolved so they don't waste energy with this biochemical process. Thus they can use the energy saved from not having to perform photorespiration to build more plant—and from an anthropogenic perspective, to make more food for us to eat. As a result, C4 plants such as maize can have yields of up to 50 percent higher than C3 plants. Tammy and Rowan and the team they work with want to engineer a rice plant to act like corn so it can make more food with the same amount of sunlight.

The problem for the research team is that it took nature somewhere between a hundred thousand and a few million years to evolve this system to begin with. (Since then, about sixty additional plant varieties have evolved the C4 anatomy.) To replicate the C4 anatomy in a plant that has evolved to work in an entirely different way is a scientific feat that will draw upon the most advanced science and the sharpest minds. Success would be an example of humans breaking through the barriers of the natural world. The project requires knowledge of photosynthesis, cell biology, enzyme analysis, developmental biology, ecophysiology, and biochemistry, as well as expertise in molecular transgenics by people known as "gene jockeys" for their ability to manipulate the double helix of DNA.

The idea of bioengineering a more productive rice was first proposed to Rowan in 1999, after he and Tammy had been in Toronto for only a few years. James Sheehy, a British agronomist at the International Rice Research Institute, held a first, exploratory meeting with some academics, and Rowan was immediately interested. But the project didn't take off for almost ten years. In 2008, Rowan and other C4 biologists travelled with Sheehy to a meeting with the Gates Foundation in Seattle to convince the philanthropic organization that this was a project worthy of its support. "We said, Mother Nature has done it so many times. If nature can do it fifty times, humans should be able to do it once." Funding came through that fall, and the C4 rice consortium was officially launched.

No one is pretending the job will be easy. The first step is to understand what inside a plant makes C4 happen. Is there possibly one genetic switch that turns on the C4 transition? Can this gene be expressed in rice? What enzymes are needed? The answers to these questions could take ten years to find, said Rowan. If successful, researchers would then require possibly a further decade or more to optimize the way the C4 anatomy works in the rice leaf. Once that is mastered, plant breeders would cross the transgenic rice plants containing the C4 structure with locally adapted varieties. At that point, decades into the future, the new rice seeds would be evaluated by food and safety organizations around the world, tested, judged, and approved—or not. And only once the seeds had passed regulatory hurdles could they be distributed to farmers.

To crack this code, then, could take fifty years and cost up to a billion dollars. But Rowan Sage doesn't balk at either the time frame or the cost. "At NASA, for one satellite to go to Mars, this costs one to two billion," he said. "A navy cruiser is close to one billion. From what society pays for, especially from the defence end, this is a drop in the bucket. We say, give us a billion dollars and we'll create one trillion dollars and feed the world's people.

"For biologists, it's our moonshot."

The puzzle these scientists are trying to solve is so big that there are many different starting points, and each research team has been delegated an avenue of investigation. At stations in Taiwan and Korea, people have already begun to look for mutant genes in rice. They are using a technique called high-throughput genomics to see if they can find a clue to how to get a plant to move from C3 to C4 photosynthesis. Researchers treated forty thousand different lines of rice seed with radiation or a carcinogenic solution to induce their DNA to spontaneously mutate and then planted the mutated seeds in experimental fields. Technicians hired by the International Rice Research Institute then examined all the rice plants in the research fields with hand-held microscopes to see whether any of the genetic mutations the scientists triggered gave the leaves any C4 properties, such as a narrowing of vein space or enlarged cells around the veins. If the technicians saw any hint of C4 characteristics in a leaf, they collected seed from the mutant plants and sent them to the headquarters in the Philippines. There, the seeds were grown into several generations of plants in order to create new lines of rice that express the mutant genes they wish to study further. Next, another team of plant breeders examined the rice lines and snipped out any intriguing pieces of DNA to send to molecular labs in Germany and England. In England, a Canadian scientist is introducing the potential C4 genes into rice plants using the tools of genetic engineering and is examining these plants for characteristics such as their photosynthetic capacity. They are then shipped to physiology labs in the United States and Australia, as well as to the Sage Lab in Toronto, where the plants will be examined further to see if they hold the clues the researchers are hunting for. "The big thing they are looking for is the transcription factor," said Tammy. "Is there a master switch in the transition from C3 to C4? Or is there more?"

This enormous network of people meet two times a year for about a week in the Philippines and have regular Skype conferences to discuss their findings and keep abreast of the progress around the world. The goal is to take all this information that is being collected in the disparate fields and laboratories and synthesize it so they can try to figure out how to generate this new kind of rice.

"Look at all these problems facing us," said Tammy. "If we don't stop carbon dioxide emissions, and considering population growth, we're going to have to have some means of people producing their food at higher yields than they are now. Rice feeds more than half the world's population. So having that capability will be profound. Because C4 plants are more nitrogen-use efficient, you will be able to minimize the amount of nitrogen you have to dump in the environment. There will be less environmental pollution. When we can't push plants any further, we have to do something from a technological point of view. That's where C4 comes in.

"But it's not just rice," she continued. "It's soybeans, wheat. If you look at the majority of our food crops, only corn is C4. The biggest crops are C3. The hope is on a really larger scale. This is not just about C4 rice but C4 soybeans, C4 wheat, C4 barley."

If C4 rice were to be released today, it would most likely be met with outrage and controversy. The new seeds, if they are ever created, will be the product of one of the most publicly contentious fields in modern science. By attempting to modify the rice genome, the consortium has entered into the high-stakes debate around biotechnology and genetically modified seeds. If C4 rice were available now, it possibly would be banned in the European Union, where legislation regulates the use of genetically modified (also known as GM or GMO) products and requires food producers to label all foods containing any genetically modified ingredients. The EU's laws governing GMOs reflect a fierce opposition to genetically modified products. When a trace amount of genetically modified material was discovered in a shipment of flax from Canada to Europe in 2009, all trade in the seeds was immediately halted. The union has a zero-tolerance policy towards GM flax, and it is not approved for human consumption in many countries.

The strict rules are also a direct reflection of the European public's sentiments towards the technology. Anti-GMO activists have broken into test fields and destroyed crops in Great Britain, Portugal, Italy, Germany, and France (where Reuters reported in 2006 that half of all test fields are damaged by activists every year). In India, where there also has been a surge of opposition against the technology, at the beginning of 2010, the government backtracked on its decision to permit a genetically modified eggplant. The environment minister who put a hold on approving the vegetable was responding to critics who feared the new eggplant, bred to resist pests without the need of chemical pesticides, would endanger the country's heritage varieties and landraces as well as rope farmers into a never-ending cycle of purchasing seed when in the past they had used their saved seed for free. Even in North America, where farmers have largely accepted the new technology and have planted millions of hectares of GM soy, sugar beet, and canola, there remains a vocal contingent that believes that genetic engineering creates "Frankenfoods" and should be outlawed. There is reason to believe that, if the C4 rice comes to fruition, those who are against the engineering of the plant genome would not see it as the panacea the consortium believes it to be and would do everything they could to stop it.

What is a GMO anyway? Biotechnology offers the tools to create genetically modified organisms. It works at the molecular level, cutting a gene from the double helix of one cell's DNA and inserting that gene into another cell's DNA. Whereas traditional plant breeding involves moving genetic traits between two closely related species, biotech allows scientists to swap genes from any species—from a fish to a tomato, from a tree to a flower. There are a couple of ways to do this. You can use an expensive, bulky gizmo called a gene gun that literally shoots a gene into another organism's nucleus; you then cross your fingers and hope the genetic material is incorporated into the DNA. The most common technique, however, is to use a kind of bacterium called the agrobacterium. In the natural world, agrobacteria live off the unwitting kindness of other species. They survive in nature by transferring their own genes into the nuclei of their host plant's cells, genes that prompt their host to grow a tumour that in turn houses and feeds the bacteria. For the past thirty-odd years, humans have harnessed this process to work on our behalf. To transfer DNA between organisms, scientists replace the bacteria's tumour-triggering gene with the gene they want to experiment with. When the bacteria infect the host cell, instead of moving their tumour-causing genes into the nuclei, they move this human-selected genetic material. Voilà! Thanks to the work of the bacteria, the cell is transformed. Then, to grow this cell into a GM plant, the scientists let the cell multiply through cell division for a few weeks, then introduce plant hormones into the cells that prompt them to grow roots and shoots. A genetically modified organism is born.

About 10 percent of the world's farmland is sowed with GM crops, mostly soy, corn, canola, and cotton. In the United States farmers grow these GM crops as well as modified squash and papaya among a few others. In Canada, a more eager GM adopter, millions of hectares of Roundup Ready canola and corn are planted, bioengineered to be resistant to the herbicide glyphosate, trademarked by Monsanto as Roundup. While Europeans and Indians have been wary of the technology (though India is nevertheless one of the top ten producers of GM crops), farmers in the highest GMO-producing countries—India, Canada, the United States, Brazil, Argentina, the Philippines, Uruguay, Paraguay, South Africa, and China—are growing an increasing amount of GM seed. In 2001, this market was estimated to be worth $3 billion. That jumped to $6 billion by 2006.[36] Although there has been some public dissent in North America, it would be safe to say that most people aren't aware that they eat food produced with genetically modified products every day. Unless certified organic, mayonnaise is made by whipping the oil pressed from GM canola, and our beef cows, chickens, and hogs feed on GM corn—and so do we when we eat a corn tortilla or nacho chips. We consume all sorts of processed foods produced with genetically modified canola, sugar beets, soybeans, and corn that is turned into high glucose corn syrup. People are similarly unaware of the large extent to which modern medicine relies on genetic modification.

Scientists have also started to apply this science to animals and fish. At the University of Guelph, researchers created what they called the "genetically enhanced" Enviropig, a Yorkshire breed of hog designed for the industrial food system. Ordinarily, pigs don't digest the phosphorus contained in grains, and excrete it in their manure. A big problem with large-scale hog production is that pigs create massive quantities of manure. When their waste makes its way into local water systems, that phosphorus feeds algal growth, which reduces oxygen levels and can kill fish. Scientists modified the Enviropig's genes to enable its salivary glands to produce phytase, an enzyme that digests phosphorus. The Enviropig's creators claimed that its manure was low in phosphorus and therefore not as polluting as that of its non-GMO cousins.

Research conducted at Memorial University in Newfoundland has been used by a Massachusetts-based company, AquaBounty Technologies, to create a genetically modified salmon that can reach market weight in sixteen to eighteen months, instead of the thirty it would take the fish if it were left up to nature. To get it to grow more quickly, scientists have added to its DNA a growth hormone gene from a chinook salmon and another piece of genetic code from an eel-like species called a pout. In 2011, news spread quickly of a cow that could produce human-like milk thanks to changes made to its DNA by researchers at the China Agricultural University in Beijing.

When I posted a link to an article about this intervention on my Facebook page, a reader summed up many people's reactions to this technology: "That's just sick."

Perhaps this yuck factor explains why genetically modified seeds are commercially years ahead of GM animals and, despite some controversy, more accepted. Even though the last Enviropig herd was euthanized in 2012, because no investors were interested in funding the project, and AquaBounty's salmon has stirred up controversy in Canada and the United States, some expect GM animals to be commercialized soon enough. As more and more countries produce increasing amounts of GM grains and food products, they are quickly becoming the status quo.

Humans have been modifying plants since we started farming, interfering with evolution by choosing to breed certain traits over others. Is manipulating the DNA of an organism simply an extension of what we have always done, or is it going too far? Let's tease out the GM debate in the context of the big question of how to feed the world in 2050.

There are some extremely accomplished and educated people who speak out in favour of the technology. Manfred Kern retired from his job as head of sustainable development at Bayer CropScience in Germany in 2011. Trained as a scientist, he devoted his career to food, researching crop protection and then leading a project called Vision 2025/2050, which looked at how to safeguard the world's food supply. For Kern, genetically modified seeds are central to a food system that is capable of feeding people in the future. "It's one of the most efficient ways to reorganize resources," he said. "Genetic engineering will add value to the genetic material. Seeds are the cornerstone." Providing people in the developing world with better seeds, he says, will increase food security. "More than 40 percent of what you are doing is related to the genes. Without water, there is no growing. Without soil, there is no growing. But the seeds are the prerequisite." Kern is most concerned about ensuring that the technology makes its way to countries in Africa and the developing world. He tips his hat to the debate, but believes that the most ethical way to proceed is to further explore the potential of the technology. "I can understand their questions. I can understand their concerns. But I think we have to try it first. We have to test the working hypothesis for the people. But to say no by definition because it is done by genetic engineering—that I can't accept. I'm not saying biotech will save the world. But it's a tool. It's an option."

The critics, however, are more cautious. For one, they fear environmental repercussions. They too look at the last ten thousand years of plant breeding and say that rather than being the next logical step, genetic engineering is an anomaly. The crop plants we consume have been tested for human safety over millennia, whereas the new biotech creations are only decades old—at most. And it's not like we're very good at assessing the risks of our new inventions. It wasn't that long ago that we sprayed our fields with DDT. Another concern is that introducing new genes into a plant could make the transformed species allergenic, which happened when a Brazil nut gene was engineered into a soybean; because of the allergenicity of those soybeans, they never made it to market. There is also a legitimate fear that government regulation of biotechnology isn't good enough. Rules don't begin to protect the public against consequences that are yet to be detected.

But the biggest problem with biotechnology isn't the science. The problem is the business and the way the seed industry uses the science. Just as concentration of ownership has reduced farmers' seed choices, an international legal structure limits what farmers can and cannot do with seeds. Intellectual property law provides a legal framework for people and companies to patent seeds and the propagating material of the new varieties they develop. Patents for seeds have existed since the 1930s in the United States. But since the 1980s, all sorts of life forms have been legally determined to be private property.

These laws exist to varying degrees in most countries. In North America, you can patent genes, micro-organisms, and new crop varieties. The European Union allows for patents on biotech materials, including genetic information, but it excludes plant varieties. Small-scale farmers in Europe are also allowed to freely collect seeds from specific plant varieties to use on their own farms as long as they don't seed too large an area.

In the rest of the world, there is a belief that life forms shouldn't be patented. People are more concerned with the rights of farmers to save, sell, and exchange their seeds and to share in their benefits. They recognize the role farmers have played over millennia in developing the genetic biodiversity of our food crops. In Latin America, you can't register a patent on plants because it is believed that the plant's genetic information belongs to all of humanity.

But international law is now forcing some countries to change their policies. The World Trade Organization requires countries to legislate what are called plant breeders' rights—these are the rights of the seed-breeding corporations to protect their financial investments. India and China amended their laws to comply with the WTO and have allowed the patenting of micro-organisms. The logic is that all this research into creating new seeds costs a lot of money. Monsanto spent at least $1 billion to create its first GM seed product. And when private companies and stockholders are footing the bill, they want a return on their investment.

These laws have a direct and profound effect on farmers. When farmers decide to grow GM seeds, they must sign a contract with the seed's producer. The companies then employ lawyers to enforce these contracts and hire inspectors who conduct on-farm investigations to make sure no one else is "pirating" their seeds. The Washington, DC–based not-for-profit Center for Food Safety, in its report "Monsanto vs. U.S. Farmers," alleged that the company uses "harassment, investigation and prosecution" to protect its intellectual property. According to the report, Monsanto has a budget of more than $10 million a year as well as a staff of about seventy-five whose only job is to investigate and prosecute farmers.[37]

One of the most famous cases is the story of a Canadian named Percy Schmeiser from Saskatchewan who grew canola for four decades, saving his seed to replant year after year. Then one year, pollen from GM canola might have blown in from a neighbour's field, or perhaps GM seed fell off a passing truck. However it happened, at least part of Schmeiser's seed stock acquired Monsanto's patented gene that made the plant resistant to Monsanto's herbicide Roundup. Schmeiser planted it. Monsanto then asked him to pay for a licence to grow the canola containing its patented genes.

According to the Center for Food Safety's report about Monsanto, many of the farmers who have been accused by the company of infringing on their patents choose to settle the dispute out of court—whether or not they are innocent of what they've been accused of. But Schmeiser fought back. Not only did he fight Monsanto in court when it sued him for patent infringement, but he filed his own lawsuit against the company, arguing that it had destroyed *his* seeds, the landraces he'd spent his lifetime cultivating.

A judge at the Federal Court ruled in Monsanto's favour, reasoning that, regardless of how the patented canola genes arrived on Schmeiser's land, all his seeds were the property of the company because in Canada, patents have precedence over farmers' rights. Schmeiser was ordered to pay Monsanto. He appealed, and the case went all the way to the Supreme Court of Canada, where in a five-to-four decision, the court upheld Monsanto's patent. However, it ruled that Schmeiser did not in fact owe the company any money in damages or in legal fees for a very technical reason: while the genes in the canola may have been protected by intellectual property law, Schmeiser did not spray his fields with Roundup, and therefore, the court reasoned, he did not reap the benefits of Monsanto's patent and did not owe the company any damages. This case underlines the serious problems with the seed industry and the international legal framework that stands in the way of better food security in the future.

This patent system and intellectual property law have fundamentally altered our relationship to seeds and to biodiversity. Whereas seeds used to be a resource farmers shared, sold, and traded with the knowledge that the free exchange of genetic material was better for everyone, intellectual property law has turned this upside down. In fact, the law stands in the way of all these acts that created our agricultural biodiversity in the first place, and it takes control of the seed stock away from the people who grow our food. It is absurd that a private corporation can register a patent on a product of nature such as a bacterium or a gene, thus restricting other people's ability to benefit from scientific discovery without paying them first. This legal framework actively stands in the way of the creation of a diverse and resilient food system.

The problem with GMOs goes beyond intellectual property law. The seeds themselves have failed to deliver and have even caused environmental harm. When you consider the possibilities of GM technology, there are countless traits that scientists could breed into crops. However, the biotech industry has focused on developing commercially only two traits: insect resistance (Bt crops) and herbicide resistance (glyphosate-resistant crops). In Bt cotton, for example, the plant's DNA is engineered to contain the gene that produces the *Bacillus thuringiensis* (or Bt) toxin that kills pests. If an insect tries to eat the plant, the insect dies. Glyphosate-resistant cotton is herbicide tolerant, which allows farmers to spray their fields with the chemical to kill weeds while the crop survives. Despite the promises of the biotech industry that their seeds would help to reduce chemical use, such gains haven't materialized. Instead, according to a report on the global seed industry by the ETC Group, an Ottawa-based international civil-society organization that examines the impact of new technology on society and the environment, farmers who plant herbicide-resistant crops have had to apply more and more herbicides in their fields as weeds developed resistance to the chemicals. And the problems caused by herbicide resistance are now exploding.

A whopping 85 percent of the world's genetically modified crops are herbicide-resistant, according to Bill Freese, a science policy analyst at the Center for Food Safety who has been working in the area of genetically modified crops since 1999. Before GM plants existed, farmers killed weeds with a handful of herbicides they sprayed on the fields before the crop sprouted so the chemicals wouldn't damage them. With glyphosate-resistant canola, though, a farmer could spray after the crop had started to grow because the canola wouldn't be harmed by the herbicide. However, thanks to evolution, the genetically engineered plants are no longer the only ones that can survive being sprayed. Weeds have no developed resistance to the herbicide. Farmers in the United States are dealing with a grave problem. According to the US Department of Agriculture, as much as seven million acres of genetically engineered soy and cotton plants are infested with these herbicide-resistant weeds. A report by Food and Water Watch puts that number at twelve million acres. In 2013, the agribusiness research consultant firm Stratus released results of their own study that tracked a dramatic rise in resistance. The total number of acres where farmers have found weeds to be unaffected by herbicides increased by 25 percent in 2011 and then by 51 percent in 2012. And on these farms, there is often more than one species farmers can no longer kill with the same chemicals.

"It's natural selection," said Freese of how weeds can evolve to become resistant to the chemical designed to kill them. Some plants have mutations that make them naturally resistant to herbicide. When a farmer sprays, the chemical kills the plants that don't have natural resistance, leaving only those with the mutant gene to pass their DNA on to the next generation. The farmer inadvertently *selects* for glyphosate resistance. The scourge of herbicide resistance has forced cotton farmers in Georgia to spend two to four times more on weed control, said Freese. They must hire labourers to weed their fields using a hand-held hoe, at a cost of between twenty-five and a hundred dollars an acre, and pay for more herbicide. To fight these super weeds, farmers are also using a chemical soup, mixing 2,4-D and arsenic-based herbicides, sometimes with the glyphosate, to try to kill them. "The Roundup Ready system is being eroded," he said. With GM crops on the rise in developing countries, herbicide resistance is predicted to increase. "They want to paint a happy face on biotech," said Freese, "but we're going to see this repeated in the future." And we're beginning to see signs that insect resistance to Bt crops is going to follow a similar path.

Even though glyphosate resistance has been well documented, there has been no move to stop producing the highly profitable seeds that have caused the problem. Rather, seed companies are moving towards what's called gene stacking. They are adding several new genes to one seed variety at the same time, so that a farmer who wants to buy seeds that are, say, bred to express Bt toxin will have no choice but to buy ones that are also herbicide-resistant.

The Union of Concerned Scientists is a not-for-profit organization founded in 1969 to conduct independent scientific research and provide science-based opinions about big issues such as nuclear energy and agriculture. It has been studying biotechnology since the early 1990s and has concluded that despite all the money invested, genetic modification has not done much more than enrich the seed companies. In its 2009 report, "Failure to Yield," the union points out that while companies have talked about breeding plants to lower the environmental burden of industrial agriculture, such as plants that use nitrogen fertilizer more efficiently, they have yet to release a single crop variety that does this. It found that even the benefits the companies say the GM seeds bring farmers are questionable. For example, GM seeds are touted as boosting yields, but the report takes a look at the numbers and concludes that those gains are small.

Doug Gurian-Sherman is a senior scientist in the union's food and environment program who wrote the report. "Some farmers do benefit," he said. "But when you average it all out, when you look at it more broadly, it is not as impressive." The data he collected showed that insect-resistant corn provided only a 3 to 4 percent increase in crop yield and that herbicide resistance as a genetic trait did nothing to increase the harvest. Similar numbers have been found elsewhere. In India, the anti-GMO organization Navdanya, founded by the internationally renowned activist Vandana Shiva, claims that the yields of Bt cotton in that country are about a third of what seed companies say they should be. The gap between advertised results and reality has even been ruled on by third-party adjudication. In 1998, the state of Mississippi's seed arbitration council, part of the Department of Agriculture and Commerce, found, in a non-binding decision, that three cotton farmers who planted GM seed and then, according to the *New York Times*, "watched as their cotton balls shrivelled and fell to the ground" should be compensated more than $1.9 million by Monsanto.[39]

"Ultimately as a society, we have to think of opportunity costs. What are the costs of going down one path and not another?" said Gurian-Sherman. "I'm not suggesting genetic engineering has no role and won't have any success. I'm saying, when you add it all up, it doesn't add up to very much."

The problems with the biotech industry seem big enough that we might want to abandon the whole GM project—that is what many people, such as Vandana Shiva, suggest we should do. Their reasoning is that the science is tainted, and besides, often the seed technology eliminates choice. After her organization, Navdanya, released a report that examined GMOs, Shiva was quoted in the *Guardian* saying, "The GM model of farming undermines farmers trying to farm ecologically. Co-existence between GM and conventional crops is not possible because genetic pollution and contamination of conventional crops is impossible to control." She makes valid points. However, if we are searching for the best tools to help us to achieve our goals of feeding the growing world population with minimal impact on the environment, then perhaps there is a way to rescue the science and the innovation from this morass.

In fact, a rescue plan is well under way in Australia. In the city of Canberra, a scientist named Richard Jefferson has created a system he believes can democratize GM technology—change it from being a tool of the rich and corporate into a science at the disposal of all human beings. Jefferson began his science career in the 1970s, working in molecular biology labs, first at the University of California, later earning a PhD at the University of Colorado. It was while he worked in the lab that he became politicized. Whereas his colleagues were quickly becoming proprietary about their research, registering patents on their discoveries, he was pulled in the other direction.

He wanted innovation in biotech to be shared in the same way that scientists had always shared their results so they could build upon one another's discoveries. So when he created an important biotech tool—a reporter gene he named GUS that helps scientists see whether a gene they want to move has been incorporated into the DNA—he not only let other scientists use it for free but gave them enough information so they could improve upon it if they wished. This kind of openness is an anomaly in the field today.

Right now, it can be hard for scientists to even create new genetically modified varieties because every aspect of the process has been patented. From the tools they use to mark the gene they want to isolate to the method they use to silence a gene so as to allow another trait to be expressed, each stage in the process belongs to someone under intellectual property law. Take the agrobacterium. To harness the powers of these bacteria and insert foreign genetic material into the DNA of another organism, you have to pay the patent owner. Similarly, if you want to use the gene gun to shoot DNA into a cell nucleus, you've got to pay Cornell University, where the gun was created. This system restricts science. The price of discovery is too high for researchers in developing countries to invest in, and even for scientists working in public or non-profit institutions. If someone discovers that a component of something they've just created already is patented, all their work could be in vain. This is why Jefferson has created an alternative to the agrobacterium he called TransBacter, which he released for free to facilitate improvements in crop breeding, reduce dependence on the multinational seed companies, and help smaller entities to innovate. He also came up with a system he calls biological open source.

Biological open source is not unlike the open source software movement. The belief is that if people share the biological tools necessary to innovate—like the TransBacter—and collaborate, we will all benefit. In a system governed by the laws of biological open source, people are allowed to freely use the tool kits that someone else has created, but in return they must share too. If the system sounds awfully familiar, it's because we know it well. "From my perspective, what we see as open source is evocative of what's happened in the last seven to eight thousand years," Jefferson explained to me. "It has been the bedrock on which we've built civilization. It's called agriculture."

Through his organization, Cambia, Jefferson has devised a number of ways to change how things are done, including creating a licensing structure called BiOS to facilitate non-proprietary sharing. Open source doesn't eliminate the opportunity for making a profit from invention. Just as in the computer industry, it simply frees the tools from restrictive regulation so that everyone has a chance to make technology—and money too. "Sharing is not just a matter of kumbaya," said Jefferson. "It is in your best interest to create a commons that nourishes. If you want to change patterns of social equity, open source provides many of the intellectual tools to rebuild on. We focus so much as a society on capturing value from the innovative process. We forget the real value to society is robust sustainable farming communities."

Can we wrestle the technology of genetic modification from the patent system? It's too early to say. Other people besides Richard Jefferson have articulated the same basic idea. Some call it open source biology, some open biology; all suggest an intellectual commons for biotech. But so far, no one has managed to achieve the critical mass required to transform the status quo. A lot of money is at stake. Universities are keen to keep the profits they gain from inventions, and corporations fear losing their competitive edge as well as money invested in research and development by sharing. But for biotech to work to the benefit of our goals for 2050, open source must be adopted. After all, a spirit of collaboration and sharing is the way the people who have bred better seeds over the millennia have always done it. Patents are a brief exception in thousands of years of working together. Judging from biotech's poor record in crop breeding so far, the only way to ensure that the science can help us all is to overhaul the patent system and the corporate control of seed breeding. "Science doesn't automatically turn into a benefit for society," Jefferson told me. "I suggest we pull back so we can remake the structure. We need revolution."

There is also the larger question of whether we need biotech in the first place. Are GM seeds reflective of the kind of society we want to live in? If you take the business and the proprietary competition out of GM seed breeding, is it good for us?

If we are going to meet the challenges of the future, we need to be what climate scientist David Lobell, assistant professor of environmental earth system science at Stanford University, calls an adaptation agnostic. Although he doesn't use the term in reference to GMOs, the concept lends itself well to the discussion. An adaptation agnostic is someone who approaches climate change solutions with an open mind and evaluates adaptations as we move forward in a way that is similar to how the medical field uses randomized trials to assess treatments. "It should be objectively clear and not based on our own convictions," he explained to me. While time is limited when it comes to seed breeding—we need to move quickly to respond to climate change and don't have twenty years to wait and see—the idea that we should free ourselves from ideology as much as possible is a useful addition to the debate about genetic modification. "We don't know what is going to work," he said to me about climate change adaptation. "We are experimenting here. By experimenting, we should be learning from what we are doing."

There could be room for biotech in our future. We have to figure that out.

As much as technology may help us, the most important first step in feeding the future is biodiversity. Without protecting this common natural heritage, we can do little with science. The debate is moot without conservation and preservation. And both the protection of biodiversity and biological open source are founded on the same principle: Our seeds belong to the biosphere. They should be for all of humanity to share. They are the only foundation we have on which to build a fairer, more sustainable food system. And that is exactly what people are working for in villages and on farms around the world.

CHAPTER TEN

SOS: Save Our Seeds

We left Yuanyang just before noon, Rose and I both hungry for food. On our way to the provincial capital, Kunming, via the brand-new empty highway rather than the back roads we had taken on our way there, we stopped for local delicacies. It turned out that Rose was a devoted foodie. In the first town we passed, she knew the best restaurant and ordered us a salad made from wild mountain greens. On the outskirts of another, she stopped the van at a small wooden hut where she bought us strawberry-flavoured water buffalo milk, sold in plastic sacks. When we passed a long row of new government-built shops selling the area's speciality, *tian bai jiu,* she took us to the one with the superior sweet fermented rice and bought a big jar to take home. And at a fruit market at the side of the road when we were descending from the lush mountains, driving out of the tropical landscape of banana trees and rice terraces into the river valley, I bought a fresh coconut that a woman drilled a hole into so I could drink its sweet water through a straw. I also bought some bananas. These bananas looked similar to the ones I buy at the supermarket back home, but they tasted unlike any banana I've ever eaten. Their perfume filled the van and their taste was of sweet flowers, with a mouthfeel that was soft and almost slippery. Ever since learning that the bananas we buy at supermarkets in the West are the same variety called Cavendish, known by connoisseurs to be one of the least flavourful, I'd been longing to taste a different kind. It wasn't surprising to find other varieties of bananas in the mountainous terrain around Yuanyang because this area of Asia is not too far from what Nikolai Vavilov, the Russian scientist who travelled the world searching for biodiversity in our food crops, would have called a centre of origin for the banana. It was once believed that the banana originated in Southeast Asia, but it was later confirmed that in fact the centre of origin for the banana reached from India east towards Malaysia and Indonesia. And it could have been in a mountainous area like this where the wild plant was domesticated.

Rose grew up in a similar area in the 1980s, isolated from the rest of the world. Her village was in a valley, with mountains visible on the horizon in every direction, cutting them off both geographically and culturally. In her community, the rhythms of life were in sync with the seasons and the farming cycle. They were affected by the Chinese agricultural policy of the day but not by popular culture or even technological change. There were no pop songs, no movies, no latest fashions. No phones, televisions, tractors, or motorcycles. There wasn't even a rice thresher to separate the grain from the chaff, and to this day her parents cut the rice by hand and then dry it in flat bamboo baskets that Rose's father weaves himself. "When I was in middle school, I thought that heaven and hell were on the other side of the mountain," Rose told me in the van on our trip back to Kunming. There was sadness in her voice, as if she felt that she had missed out on something as a child, on the deeper understanding of the world that she has gained as an adult, now that she lives in Kunming and works as a guide. "When you go out, you know about China, and the globe. You learn about the whole world." Rose was raised to be a farmer. Although she spent most of the year at the boarding school in town—that seven-kilometre walk separating her from home—she did return to live with her parents twice a year, during both winter and summer vacation. It was on these extended stays that Rose helped with the chores and learned from her parents how to grow rice.

Rose's parents likely knew nothing of Vavilov's theories of biodiversity and his enormous collection of seeds. Nevertheless, they knew—like all farmers—the value of these small grains of life, and they taught Rose this appreciation. They knew that the quality of their crops, the yields and the hardiness, depended on what they sowed in the earth at the beginning of the season. And just like Vavilov found on his expeditions, interviewing peasant farmers around the world and collecting their seeds, Rose's mom and dad were aware that their seeds were improved by mixing, by diversity. "My parents say when you exchange your seeds with the neighbour, it's much better—the more yield you'll have. So you must exchange every two years."

The seeds, selected by the family for the vigour of their parent plants, were first sewn in wet soil near their house in the village. Rose was taught to put two seeds in each hole she made in the wet earth to ensure that they'd have enough seedlings if some didn't germinate. By starting their rice plants near their home, the family gained more time to let their wheat mature in the terraces and harvest it before transplanting the next season's rice crop there. After the wheat was collected and the rice seedlings had sprouted, they prepared the terraces for the rice. Because Rose's parents drained the area so the wheat could grow in the dry season (wheat doesn't need as much water as rice), the first step to readying their fields for rice was to flood the paddies. At the beginning of the dry season, they had dammed the river that flowed above their land with stones; now it was time to open these stones and allow the water to flow into the terraces so the soil could become saturated. Then her parents guided a water buffalo into the plots so the animal's big, heavy feet mixed the water with the soil to make a thick, soupy mud. It was then the children's responsibility to transplant the seedlings, using their small hands to nestle the roots into the muddy earth, spacing them about ten centimetres apart.

The family kept a close watch on their terraces to make sure there was always enough water for the rice and devised their own systems to fertilize their plots and to keep the weeds down. Typically, to remove the weeds in the traditional rice terrace requires hours of stooping over to pull out the unwanted grasses by hand. But Rose's father knew these grasses could be used to his benefit. "My daddy is very clever," she said. "My daddy's way is to use the feet to dig a hole and dig the grass into the mud so it becomes the fertilizer." Another way he boosted fertility was to shovel pig and ox manure into the water up by the stone dam so that gravity dispersed the nitrogen-enriching matter throughout the terraces. Nevertheless, their lives as small-scale farmers were still full of a lot of hard work. "My parents don't have spare time," Rose said. "When I was growing up, the most fun we had together was playing cards. My mother is very happy when we play cards with her."

Then one day, the outside world came to Rose's village. A seed company rented some of the terraces and planted new high-yielding varieties "to show what hybrid rice looks like to the villagers," said Rose. People watched. In the evening in the village, it is a custom for neighbours to drop in on one another and to chat. During these visits, there must have been many discussions about this peculiar new rice that seemed to be doing very well—in fact, much better than their own—in the rented fields. "The village members, they talked about it between themselves," remembered Rose. They were impressed. "Then they start to try it." First one family planted some of the new seeds, then another and another. Over a period of five years, the whole village switched to planting the new rice varieties, purchasing their seeds from the company rather than saving their old seed lines. Everyone, that is, except for the poorest family, who lived in the most mountainous area and couldn't afford to pay for the new seeds.

Rose's parents proceeded with caution. The first year, they planted the new seeds on only a small portion of their land—a third of a mu, a Chinese unit of measure equivalent to 666 square metres—to see whether the seeds would perform. They followed the instructions of the seed company and purchased the chemical fertilizers, small pellets that came in a sack, as well as pesticides to spray on the insects that preyed on the rice. That first season, they were pleased with their yields, and the next year they planted even more of the new variety, forgoing some traditional rice. Yet they didn't give up all of their seeds. In a separate terrace, they continued to plant their generations-old sticky rice so they could still make special dishes for celebrations like the Winter Solstice Festival.

The story of Rose's family's move from traditional varieties to mostly growing newly developed seeds follows the bigger-picture narrative of what has happened to seeds in small farming villages around the world.

Since the late 1960s, this same scenario has been played out across Asia, as farmers abandoned their traditional rice farming for green revolution–style agriculture. In countries such as Vietnam, Thailand, South Korea, Bangladesh, and Indonesia, small-scale farmers like Rose's family have been encouraged by their governments and agribusiness to modernize and adopt new seeds, new fertilizers, new methods of pest control, and new irrigation systems. This transition is what Ivette Perfecto, a professor at the University of Michigan who studies biodiversity in agriculture, calls technification. And it's a pattern that has been repeated around the world, with a range of crops. In Latin America, for example, coffee farmers were encouraged to intensify their production and were instructed to do so by eliminating the shade trees that shelter coffee bushes in traditional systems.

This agricultural shift involves a total transformation of the way farming is practised, and at its base are the seeds. The new rice seeds that were distributed and sold in Asia's green revolution primarily came out of research done at the International Rice Research Institute in the Philippines. (China and Japan also have had their own seed-breeding programs.) In the Philippines, scientists were inspired by the work done with wheat in Mexico and wanted to replicate that success with rice. As with all green revolution efforts, it was a modernization project during the Cold War. The institute consisted of an 80-hectare experimental farm in the municipality of Los Baños. Different soil types were brought in from all over Asia to mimic various growing conditions in the plots, which even had irrigation pipes underground to ensure a perfect growing environment. At the institute, there were also fancy American-style homes where the scientists could live in modern luxury. It was such a sight to behold that people came on bus tours to gawk at what was perceived as Western progress in their country. In the 1960s scientists there created a variety of rice called IR8, dubbed Miracle Rice. The seeds were launched with much fanfare, even though it has since been reported that the scientists were concerned about their susceptibility to disease. Nevertheless, people went crazy for them. The rice seeds were even sold in bank lobbies and fashionable department stores.[40] Farmers across the region quickly adopted these new seeds, and scientists continued to breed even more rice lines, distributing those, too, to farmers. In China, hybrid seeds became commercially available in 1976 and spread across the country, introduced to rice-farming communities like Rose's by both government research stations and seed companies. Today, hybrid rice varieties are grown in about half of China's rice-growing areas. What this has meant, however, is a narrowing of the rice gene pool.

According to a report put out by Grain, an independent non-profit organization that works with small farmers around the world, in China over the past forty years, there has been a forty-six-fold reduction in the number of locally adapted rice varieties grown by farmers. It is the same situation across Asia. In the Philippines, 98 percent of the rice fields were planted with high-yielding varieties by the 1980s. In Thailand and Burma, five varieties account for 40 percent of rice crops, and in India, fewer and fewer of the old rices are grown today. Over the same period, however, it became more difficult for farmers in Asia to support their families. For example, in Thailand, the world's leading exporter of the grain, farmers are struggling to make a living growing the high-yielding varieties for the international market. In 2006, La Via Campesina and an Indonesian peasant federation organized an international meeting in Jakarta of activists concerned about the future of rice. The conference report described the difficult situation Thai rice farmers have found themselves in, having to borrow money from local lenders to pay for seeds and inputs. These farmers are sinking further into debt, promising to pay back local moneylenders in rice, at the same time as the price of the crop on the international commodity markets rises. Because of the way rice is sold, they are unable to capture the benefit from the price hikes. The report also described a similar situation in Sri Lanka after the World Bank recommended that the country start to produce crops for export and import rice for the local population rather than grow it in their traditional paddy systems. After years of difficulty, some farmers are now finding some relief by planting traditional rice seeds that don't require costly pesticides. The farmers are also starting to set up direct marketing plans, selling their rice through trade unions in urban areas.

It is for these reasons that a campaign in Asia is seeking to preserve the old rice lines. Save Our Rice believes that rice is more than a commodity. Rice is life and culture. It's the food upon which societies have been built. Save Our Rice believes it should be recognized as such by governments in the region and is working to preserve what's left of the old ways. This campaign is part of the worldwide sustainable food movement that is helping to build new food systems for this century.

Kunming, the Chinese city where my journey with Rose began—and where it ended—also happens to be the Chinese headquarters for a small NGO that is one of the leaders in the rice-preservation movement. The Pesticide Eco-Alternatives Center, an environmental group with offices around the world, is responding to the ecological damage it has witnessed in the paddy fields caused by the technification of rice. The intensive farming methods that the group is concerned about have been adopted to boost rice production so that the countryside's dwindling number of farmers can continue to feed China's rapidly growing cities—such as Kunming.

In the streets of Kunming, signs of these sudden changes were everywhere. To start, the traffic was awful. The city's network of wide arterial roads seemed to be in a state of constant impasse; the only vehicles moving were those in the bike lanes, where motorbikes, mopeds, and bicycles of all sorts, piled with people and cargo, zipped by the stationary cars. It was unclear whether the traffic jams were caused by too many vehicles on the road, the ubiquitous construction, or a combination of the two. Wherever you turned, workers were building fly-overs; a new urban rail line was under construction; and apartment towers—of which there were already dozens as far as the eye could see, with rows of solar water heaters on their roofs—were going up all around. The city had grown so quickly that it was fast eating up the farmers' fields on its outskirts. As we passed an expanse of rubble, a mass of bricks and concrete covering a few empty hectares beside the road, Rose explained that this detritus was the final remains of old farming villages that were being replaced with apartment blocks, towers to house people like herself who have come to the city for work. Rose's sister lives in one of these buildings. In fact, her sister married a Kunming-area farmer whose land was taken over by urban sprawl, the generational family home knocked down and their fields prepared for a building site. But Rose said his family was paid handsomely for their land and were given an apartment unit in a tower built on what used to be their fields as well as two more they rent to replace their lost farming income. "They used to not have a toilet. Now they have a Western-style toilet off their bedroom and a Chinese-style"—or squat toilet—"too!" she said, laughing. I got the sense she was a little envious.

It was in one of these apartment towers that we found the Eco-Alternatives' offices and a handful of young people at work at their desks. A few of them had recently returned from Terra Madre, the international Slow Food conference held in Italy, where they had shared the story of the work they were doing in a village called Heier in the Qujing prefecture, a few hours' drive from Kunming. In the village, farmers belonging to the Zhuang minority group grow a revered type of sticky rice named after the village. Their rice has been famous for at least several centuries, and legend has it that during the Qing dynasty, about 350 years ago, the emperor Kangxi designated Heier sticky rice his official rice because it was so flavourful. Stories like this one and the names of some of the old varieties—such as Zhefang Royal Rice—conjure up an interesting past for China's traditional rice.

However, compared with the high-producing hybrid rice varieties that have been promoted by the government in the area, Heier sticky rice doesn't yield very well. Whereas hybrid rice produces upwards of eight hundred kilograms a mu, the old-fashioned stuff provides only about three hundred kilograms. "Less and less people plant it," explained one of the women in the office. "Farmers prefer to plant the hybrid rice. This village had forty rice varieties. Now there are only six." If the farmers wanted to eat sticky rice, they could buy some grown in northwest China for less than it cost them to grow their own varieties. The reason behind this state-led push towards the new varieties, she told me, was at face value a rational one. "The government introduced people to hybrid rice years ago so they could feed everybody," she said. "For the elder Chinese, they have very bad memories of hunger."

But these high yields have come at a cost. "People find when they use hybrid rice, they have to use a lot of chemical fertilizers and pesticides," she explained. Added another woman, "Traditional rice doesn't like chemicals. If you put chemical fertilizer on this rice, the yield will be lower." One reason for this is that if you fertilize traditional varieties with nitrogen to increase yields, the plants tend to topple over under the weight of the larger seed heads. But there are benefits to growing chemical-free that can't be measured in crop yields. In a rice field where traditional seeds are grown, there exists a whole lot of life, a whole lot of biodiversity. In the water live zooplankton and nematodes and molluscs as well as surface-dwelling insects. Because rice is grown in a wetland, you will also find amphibians, reptiles, fish, and water birds all thriving amid the growing grains, as well as other vegetation. This life becomes food for humans, and the biodiversity offers the farmers the protection of ecological resilience—if one part of the ecosystem doesn't do well one year, another is sure to flourish. Compare this with the hybrid rice paddy where chemicals kill all these other life forms and turn the growing area into a monoculture. And monocultures are the opposite of resilient. Research conducted by academics at the Yunnan Agricultural University found that hybrid-rice monocultures were in fact providing the opportunity for voracious pests such as planthoppers to thrive because the ecology of the wetland is thrown out of balance by the chemicals. The pests' natural predators are killed, and the planthoppers that are resistant to the pesticide multiply in a way that they normally wouldn't have the opportunity to do. Then the planthopper destroys the crop.

So the anti-pesticide organization in Kunming is trying to encourage the farmers to keep the old ways and, if they've already made the switch, to go back to tradition. They've held workshops where they educate villagers about the health hazards of handling the agricultural chemicals and highlight the financial burden of a system in which you must purchase seeds and fertilizers with cash rather than drawing on nature for these inputs. They also make the cultural case to keep growing the sticky rice. Because the region where the village is located remains populated by ethnic minorities, people's identities are tied to rice. The Heier sticky rice is typically grown in various colours and is cooked to celebrate seasonal festivals, something that can't be replaced by imports. Also, when it comes to taste, they say nothing compares to the perfume of the local varieties.

The challenge in convincing farmers is economic. Because the yields for hybrid rice are higher, the farmers believe they will make more money growing the new seeds. "Each family believes that more yields will mean more income. But that's not true," said one of the women in Kunming. The cost of the chemical inputs and the seeds typically eats up any extra profit. Also, the Heier farmers are like so many of Asia's rice growers: they are poor with few resources. They can't afford to let concerns they may have about handling agrochemicals interfere with feeding their families. Even when the higher profits never materialize, the promise of earning more is always there, and the farmers continue to chase an elusive dream. So the Eco-Alternatives group came up with a solution: it would try to connect the farmers who grew organic, traditional rice with people in the city who were willing to pay more for a higher-quality product. "We brought urban consumers to see how farmers plant traditional rice. They see how they don't use agrochemicals and they want to pay more," said the woman in the office. "Especially older consumers. When they were young, there were no agrochemicals. Many years ago, they ate food without them and they remembered the taste. They can identify the tradition."

Since 2007, at harvest time in May and October the Eco-Alternatives group has invited people from Kunming to the village. They are able to see for themselves that the farmers are using organic methods of agriculture, and then they sit down with the farmers and help to make a planting plan based on the varieties the consumers prefer. Once the rice is harvested, the city people pay a higher price for the crop. "The consumer doesn't worry about food safety and the farmer doesn't worry about sales." The group also found that the appeal of traditional rice is not limited to the older generation or to people who make a lot of money. "For some consumers, they don't have a high economic level, but they want to eat healthy food. So one day a month they can eat this rice. The wealthier can eat it five days a week." The project is small—a mere twelve farmers were participating when I visited, which is minuscule from a Chinese perspective. But the group is hopeful it will grow. "When we encourage the farmers to harvest the rice, they harvest more and more." And save their seeds.

There's an area in Nepal between the flatlands that span the border with India and the high Himalayan mountains that is known as the Middle Hills. The principal city is Hetauda. There are few roads in the area, the steep mountain slopes making them almost impossible to construct, and people get around mostly on foot. In the summer, the Middle Hills are a scenic expanse of sloping green, but wet weather and rain make the grade impassable for several months each year, the roads frequently cut off by flash floods and landslides. It is in this part of the country where most Nepalese live.

Nepal is primarily an agricultural community; nearly two-thirds of the more than thirty million people who live there are farmers. In the Middle Hills, people grow corn, rice, and legumes, both to feed themselves and to sell to earn a bit of money. In the dry season, the farmers plant the corn in their fields and then, after harvest when the monsoons arrive and the weather is wet, they plant the rice in the same soil. Yet these people are poor. Even though they grow their own food, on the whole they don't have food security, and in 2008, the perilous state of their food system became all too evident when the corn crop failed. "There wasn't a harvest," said Kate Green, a development worker and the program manager for the Nepalese projects at USC Canada, the country's oldest humanitarian organization, founded in 1945 as the Unitarian Service Committee. Green has travelled extensively in the region, to support local NGOs that are promoting biodiversity and, in particular, that are working with preserving heritage-seed systems. She visited the Middle Hills in early 2009, several months after the harvest failed. "I remember people saying there was nothing there. There weren't kernels and the cobs didn't form all the way." What happened that year meant disaster for farmers.

In the Middle Hills of Nepal, people have been cultivating rice and corn for centuries. According to Dinesh Shrestha, head of an organization called Parivartan Nepal who has worked in this area for many years, there are hundreds of varieties of corn—red corn, yellow corn, whites, black—and tens of different types of rice. In the 1970s, hybrid corn varieties arrived, promising their higher yields, and people abandoned the traditional corns, as well as the rices. More and more farmers would buy their seeds along with the necessary agricultural inputs, and the old ways of saving, selecting, and trading seeds began to disappear. But the new seeds didn't bring the prosperity they had hoped for. Rather, the communities faced the same cycle of debt that so many farmers elsewhere were living. "They faced great problems," explained Shrestha. "The farmers tried suicide because they couldn't pay their loans."

When the crop failed at the end of 2008 and into the first months of 2009, thousands of peasants were left with nothing. Angry farmers blocked the roads and blockaded government buildings, demanding compensation for their losses.

Some of them blamed the multinational companies that sell seed in the country. The possibility was raised that their crop failure was somehow related to the collapse of the South African harvest, when GM corn similarly failed to grow kernels around the same time. However, the American and Indian seed manufacturers blamed the unusually cold weather in the region that year. Many people also pointed out that the seed system in Nepal was highly unregulated and bad seed could have easily been brought into the country by middlemen. The government launched an investigation.

No one knows for certain why the crop failed; no one knows what the government's investigation found, but local media reported that it was inconclusive. Kate Green in Ottawa hypothesizes that it could have been bad seed, possibly an old bag of stock that was no longer viable, or a sterile hybrid, or even a faulty genetically modified product that was brought across the border from India. "By looking at a corn kernel, you cannot tell its provenance," she explained. "You can't tell if it's a hybrid, if it's a GMO, or if it is a locally adapted heritage variety. Visually, there is no way to tell these things about a seed." Regardless of the cause of the crop failure, it has led to a change in attitude in the region and beyond. There is now a growing understanding of the importance of maintaining a local, farmer-based seed system to help guard against this kind of crop failure. "Now the farmers are realizing they want to have a strong control of our seed system," said Shrestha. "If we buy the seeds from outside all the time, we won't have our own control of our seeds. We've convinced our farmers that our seed is more viable in our ecology. Our own seed is more reliable."

According to Green, "The more farming communities can save their own seeds locally, they are safer and have more security in that system for good production, stable production, and even an increase in production over time in their local environment." This perspective is spreading. Since the crop failure, the government of Nepal has opened several community-run seed banks, and NGOs continue to lead the way in preserving landraces, running about a hundred seed banks in all. Shrestha's organization continues to teach the farmers to collect, save, and replant their seeds every year, instead of having to buy them. It is now working with more than ten thousand farmers. But to save the old lines will take work. "Many of our seeds have already disappeared. Our soil is getting worse every year," Shrestha told me. "But we don't want to replace our system with the so-called modern agricultural system. We need more technical support, we need support for development. Maybe support in building infrastructure for a seed system so we can improve on our own. But we don't want to import your seeds."

To wrestle our seed system away from corporate control and restore the ability of farmers around the world to propagate and plant the seeds they choose, we must support grassroots efforts with policy. Governments have a central role to play in creating the legal environment that encourages the best seeds to be bred and shared. New rules will help not just the seed activists working in the villages but also the scientists investing their time in the lab. Seeds and their genes are part of the biosphere; seeds don't belong to humanity, particularly not to a corporate entity.

Can these civil-society initiatives such as the Save Our Rice campaign coexist with projects like C4 Rice? They must. In a world where biological open source is the legal norm, there is opportunity for farmers and scientists to collaborate. They can share genetic material preserved by farmers and work together to breed the seeds we need for a future under climate change. They can take inspiration from the codependent web of life that is biodiversity on our planet.

Combine these seeds with sustainable farming, as well as an alternative economics of food that supports a new generation of farmers in addition to farmland preservation, and we will be well on our way to unspooling the industrial food system. However, to support these profound changes, we too need to change. If, as the saying goes, "you are what you eat," then we as individuals and as societies must transform too so our food culture supports a new way. This is the next and final step.

PART THREE

TARGET 2040:

CULTURE

CHAPTER ELEVEN

From Home-Cooked to Takeout: A Culture of Food for the Future

The herd was audible long before it came into sight. The sharp ring of the cowbells and the animals' guttural bellows grew louder and louder. I stood with a group of about twenty other people at a fork in a country road in the Aubrac mountains, a remote and little-known area in the southwest of France, several hours' drive from the city of Toulouse.

We had all come to the tiny village of Les Clauzels on a sunny Saturday morning in May to participate in the transhumance, the annual spring migration of the local cow herds from their winter quarters to the green pastures where they'd spend the next several months eating fresh grass. We'd each been given a stick carved from a thin but sturdy sapling, which I'd assumed was to help with the climb up the mountain, but as I stood there listening to the sound of the cows growing nearer, the eleven-year-old granddaughter of my host, Monsieur Valadier, explained to me that the stick was actually to be used to herd the cows and to help stop them from trampling us. Our job was to stand in front of a patch of green grass—grass so green and luscious that it had to be the most delicious-looking grass from the perspective of a cow that had eaten dried hay all winter—and make sure none of the animals pushed through to graze there. We were to coax the cows to pass us rather than storm the grass, so they would round the corner and continue several kilometres up the mountain to their pastures. "If you're scared, stand behind the rock," someone said, pointing to an old stone cross that was about as tall as me and, planted in the earth at this fork in the road, looked like a tombstone.

Then suddenly, the bells became louder, the ringing more and more frenetic. I could hear the farmers who were with the cows shouting to the animals. The sound of dozens of hoofs hitting the road rippled towards us like thunder, and then in a flash the herd burst around the corner. Forty large beasts with large curved horns ran directly towards our group standing in front of that succulent grass. I banged my thin little stick as hard as I could and hoped the cows would make that turn.

Which of course they did, as they had done for centuries before. And as soon as they rounded the corner, the eleven-year-old sprinted after them, taking up the rear to make sure none of the small calves were left behind. The rest of us stick bangers followed, jogging to keep up. The herd kept a steady pace, and we followed the old country road that would eventually lead us to the fields.

It wasn't long before my host, one of the local farmers, a Monsieur André Valadier, pulled up behind us in an old white pickup truck. On the bed of the vehicle was a cage of sorts to be used to transport the calves that had a hard time keeping up with their mothers—or for any of the humans in the group, some of whom had come from as far as Paris (and me from Toronto) to participate in the transhumance and weren't accustomed to the hard labour of working on a farm.

At seventy-eight years of age, Monsieur Valadier was better suited to driving the truck than running up the hill with the cows. He'd spent the last seven decades of his life accompanying his family's animals on the spring ritual, leading the excited cows towards their pastures, tending to the young ones that were hurt or left behind. Besides, the young men in the group seemed to really enjoy their calf-watching duties. When they had to catch calves that had trailed away to nibble some tasty grass, the men would go right up to the animals and try to grab them by their heads, controlling the calves with amazing force as they bucked and fought to stay and eat some more, rather than be forced onto the truck bed and be taken to their mothers. Monsieur Valadier, a tall man with broad shoulders, grey hair, and a big moustachioed smile, beamed from behind the wheel as he oversaw this procession of friends, family, and cows.

I have been on a first-name basis with many of the people I met while writing this book, but I never called Monsieur Valadier by anything but that respectful form. In France, social customs verging on old-fashioned still prevail. Someone my age would never call a person who was decades older anything less formal than monsieur or madame. So, Monsieur Valadier remained a monsieur to me, despite the time I spent with him that week as I followed him around and learned about his life and his community.

And over the course of my visit, I got to know that it's not just age that has earned Monsieur his title. Monsieur André Valadier is a hero in his community, a celebrity of sorts. Everywhere we went, both the young and the old knew who he was, and wherever we visited, he was greeted by smiling people who thrust their hands into his and pumped his arm. I heard one official call him the "Pope of Aligot"—aligot being a dish of potatoes and cheese made with the milk of the local cows. He has also been awarded France's Legion of Honour. This was all because, over his lifetime, he has led a movement to save the Aubrac's farming traditions from evisceration. In the face of pressure to conform to mainstream values and industrialize agriculture in the area, he spearheaded the building of a whole new food system that continues to grow and nourish the people of the Aubrac through the preservation of the natural world where they farm, as well as their culture. And it's this culture—the terroir of the Aubrac—that is the thread that holds everything together: their regenerative way of farming, the cows, the cheese making, their language, the native grasses, and the biodiversity of the region.

"The Aubrac is a corner of this planet where humans were placed by destiny. Here, we created means for our survival," said Monsieur Valadier. "Unfortunately, too many people who live in a similar landscape feel that, where they live, there is nothing to do to make a living. That they should leave and go elsewhere. And yet the terroir enables us to create a resource right where we are."

A poster printed by a local organization to celebrate the region sums up the defiant spirit that defines the people of the Aubrac and their agriculture. It features a photograph of the handsome face of an Aubrac cow, with its wide-reaching horns and doe eyes. On the poster are the words: "The Aubrac, the breed for a country of resistors."

The mountains in the Aubrac are the most gentle of mountains. Their tops are rounded and smooth, worn down like giant pebbles. Stand at the summit and the vista is a wide expanse of space, one slope melting into the next. The horizon is so vast and curving that it appeared to me as if someone had outstretched an arm while holding a paint brush, twirled slowly, and had drawn the outline of the mountains on the sky. The morning of the transhumance, this sky was a rich blue and these mountains were a brilliant green, dotted with dozens of different wildflowers. Delicate white narcissus, purple violets, yellow arnica blossoms, and the green stalks of *gentiane* that would soon bloom and whose roots are used to make a traditional bitter-tasting liqueur. The sun was strong, the air warm, and the crickets were so loud, they were almost deafening. Swallows swooped over the fields, catching insects and singing. It seemed like a perfect day for a walk in the mountains.

But the weather remained a concern for the farmers. It had been unusually warm that spring and it felt more like July than May, said Monsieur Valadier. The year had started out strangely, with little of the snow that typically blankets the mountains in the winter. Then when spring arrived, it came early and hot, with only some rain, and by mid-May a drought had set in. Officials were warning that water-conservation measures might soon have to be adopted. As I followed the cows up the mountain that day, I chatted with a friend of Monsieur Valadier's who had lived his whole life in the region, raising his daughters in a nearby town. He observed that the farmers had already taken in their first crop of hay, something they would typically do a month later. "Normally, when we come here for the transhumance, the grass is high," he said, looking out at the fields we were passing. Then he puffed, jogging a bit to keep apace with the cows. "They climb quickly. It's faster than the Tour de France!"

Monsieur Valadier similarly had his eye on the weather. When another guest had arrived at the house that morning and remarked that the forecast was calling for rain, Valadier smiled sheepishly and said, "I almost wish it would rain." Though he wanted to stay dry for the big hike up the mountain that day, he needed rain to water the fields where his cows would graze. "I'm scared for Jean and Géraud," said Monsieur Valadier, referring to his two grown sons who have taken the lead on the family dairy farm since he retired. "If it doesn't rain, they will have to buy all that hay for all those cows." He shook his head.

Without water, there is no grass. Without grass, there is no milk. Without milk, there is no cheese. And cheese is the base of the terroir of the Aubrac—the cheese that makes the aligot, the dish to which Monsieur Valadier is Pope.

To make aligot, you purée potatoes and add crème fraîche and then an unaged type of cheese called a tomme. You stir and stir and stir the mixture until it becomes a thick, goopy paste that, when served, must be shaken quite forcefully from the serving spoon and lands on your plate with a thud. It might appear to be a simple dish, but in fact it is quite hard to make right. A certain elastic texture is desirable, and the way you test to see if you've achieved this consistency is by stretching the paste with a large paddle-like spoon to make a long thread; in French they say "*on file l'aligot,*" which translates as "one pulls the aligot." And they do just that, stretching it into a long string like taffy. When Monsieur Valadier made the aligot, he stretched that cheese thread from the pot on the floor high into the air until it was more than six feet long. So it is fitting that this sturdy aligot thread is a metaphor for the thread that offers the Aubrac its resiliency, keeping the community together, a culture alive and a food system intact. Wherever we live, we can all find our own version of aligot—that something that can tie us together with the land, with one another, with our food, to help support a sustainable new way.

Monsieur Valadier was born in 1933, the only child of a farming couple in the tiny mountain village of Les Clauzels. If you ask Monsieur Valadier how old he is, he is likely to grin and joke that he's three hundred. Not because he feels old but because for the first part of his life, he lived as his grandparents lived—and their grandparents and their grandparents before that. He is a *paysan,* a peasant, or in the words of the local language, Occitan, which he grew up speaking at home, *lous paysan.* What he means when he uses this term is that the family lived off their own labour. They were subsistence farmers. Like all the families in their mountain community, they grew vegetables in their garden and kept chickens and ducks and turkeys as well as pigs for meat and twelve cows from which they drew their milk and made their cheese and butter. He said they rarely purchased anything, not even clothing—his mother spun the wool to knit him his sweaters and socks. What they did need to buy—sugar, chocolate, coffee, oil—they traded for with their eggs and butter. To get around, they used horses, the animals pulling sleds in the winter and wagons in the summer. The cows they raised belonged to the Aubrac breed, a genetic line selected by Monsieur's ancestors to be providers of milk but also of muscle: the animals were used for labouring too. They ate beef only when a cow died accidentally, and then they called all the neighbours together to share the food. It was an isolated life: he went to school in the village, lived amid his cousins and neighbours. The small hamlets in the area were connected one to the next by the postman, who would walk twenty kilometres a day carrying the mail but also bringing with him news by oral post: someone had died, a baby was born, there would be a marriage.

As with all small, subsistence-farming communities, life was governed by the seasons. In the winter, people remained in their villages, tending to their cows, which they would tie up in a barn. Then in the spring, the villagers would lead their cattle up the mountains as part of the transhumance, and a handful of men from the village would remain there to tend to the animals and make cheese all summer to preserve the milk that flowed during the warm season. The cheese would then sustain them in the winter. These cheese-making men—the *buronniers*—lived during the summers in small stone houses called *burons* built on the hills where they would work hard milking the cows, curdling the milk, and then pressing the fresh tomme into big cylindrical blocks to age. They called their cheese Laguiole—pronounced "la-yole"—after a village in the area. Or perhaps the village was named after the cheese, Monsieur didn't know. It's a raw-milk cheese that is aged until it develops a thick rind. When you cut open a block, the cheese inside is a creamy yellow, the pungency of its flavour dependent on how long that particular piece has aged. It is with the tomme of Laguiole—the fresh curd that hasn't been salted or aged—that they make the famous aligot.

Life remained constant in the Aubrac, said Monsieur Valadier, until after the Second World War, when an appetite for change crept into town. The desire to modernize and abandon the old ways, to buy a tractor and join industrialization's revolution that was taking agriculture into the future in countries around the world, held just as much allure in the picturesque Aubrac as it did in Saskatchewan and Iowa. At this point, the story of Monsieur Valadier and the farmers in the Aubrac looks just like the larger narrative of the food system in industrializing countries around the developing world at the time. But what makes theirs different is what happened next. While the rest of us continued on the path to further industrialize the food system, transforming every link in the food chain from the farm all the way to the consumer, Monsieur Valadier and his colleagues in the Aubrac did not. After starting down this path, they decided that it didn't feel right, that industrial farming didn't suit their culture, their way of life. And what they did next changed everything.

So what is different about the story of the farmers in the Aubrac and their Laguiole cheese is that they are about three decades ahead of the rest of us. What we're experiencing now—this social movement of people around the world who are building alternative food structures to feed the population without sapping the earth's resources—they've been there, done that. They are the future. They are *our* future. From their story of splitting off from the mainstream—their agricultural schism, if you will—we can learn what kind of a culture we need to support a new, just food order that can at least try to cope with the curveball of climate change without worsening the situation. They did it in thirty years, and so can we. Monsieur Valadier doesn't get called the Pope of Aligot for nothing.

For millennia, we have engaged with food in a uniquely human way. We roast, fry, bake, simmer, boil, grill. We knead, we roll, we wrap and chop and stir and mash. We've invented all sorts of ingenious ways to turn the natural world into delicious edibles, even harnessing the powers of other organisms such as bacteria and yeast to ferment, rise, and curdle to make our vinegars, wines, breads, and cheeses. We've come up with countless ways of turning a combination of basic ingredients into thousands, maybe millions, of dishes, each drawing on the terroir of the area where the food is rooted. We can turn grains of rice into long slippery noodles, soybeans into silky, quivering tofu, and milk from cows, sheep, goats, even camels and donkeys into creams, cheeses, butters, and yogurts. It's magic! We've dedicated so much of our time not only to preparing but also to growing and raising the foods we eat. We tend to delicate saffron crocuses. We heap earth over asparagus plants to create the more delicate tasting white spears. We dream up all sorts of ingenious ways to produce our food, such as raising fish in rice paddies like Rose's parents in Yunnan.

Of these activities, we have created métiers: vegetable farmer, orchardist, cattle rancher. Cook, baker, patisserie chef, chocolatier, chaiwala, sushi master, tofu maker, churrero, halwei—the list goes on and on. No matter where we live in the world, food is an obsession. American food writer Elisabeth Rozin captures the universal dedication humankind has for cooking: "In the dusty streets of rural villages, in the dingy rooms of city tenements, in the furtive clearings of sweating jungles, in the secret, sacred precincts of three-star restaurants, it is going on. Listen, and you will hear it: the clatter of pans, the slapping of dough, the pounding of grain. Sniff, and you will smell it: the roasting meat, the newly baked bread, the aromatic sauce. Look, and you will see it: the quick stirring of a pot, the delicate folding of a triangle of dough."

And yet, by all accounts, the cultures of food are eroding. A trip to the Western supermarket will confirm that our desire to engage with our food has diminished to the point of almost disappearing. Grab a cart and wheel over to the deli counter where there is a selection of meats cured in a factory with industrial preservatives, as well as salads—potato, macaroni, fluorescent green coleslaw—made in some complex who knows where or when. They are available today for a mere twelve dollars, to take home for dinner with a pre-roasted chicken (seasoned with salt, MSG, and the anti-caking agent silicon dioxide). Or pick up a bag of pre-washed salad mix with its own pouch of salad dressing. Not too long ago—sometime in my teens—you had to actually *wash* lettuce and roast the chicken yourself. But there's no need to make dinner tonight. Or breakfast, for that matter. Scoot over to the cereal aisle where there are boxes of grains processed into dozens of shapes and flavours as well as oatmeal that comes pre-seasoned in a bag to which you only need to add boiling water. Then there are giant muffins, and smoothies blended in a personal-sized plastic container that you merely need to open and drink. And if you prefer to eat organic, then pay a little more and the smoothies come made with organic fruit. Oh, and there are organic pop tarts too.

If you are feeling more ambitious and like to cook, at least a bit, you can find muffin mix that comes in a plastic cylinder that only requires a bit of water and a few shakes before the batter is ready to pour into the pan. There are potatoes peeled and chopped in the freezer aisle that only need to be opened and steamed so you don't have to bother with the messy peeling and cutting if you want mash with your dinner. Or perhaps you'd like to choose one of a handful of ethnic-themed frozen vegetable medleys—there's Italian pasta and Chinese stir-fry, among others.

When we buy these food products, we choose price and convenience over taste and nutrition. For good reason. We tell ourselves we have little time in our busy lives to prepare meals. For the record, I am no food puritan. I've eaten this stuff before—okay, some of it. Though the more I learn about the industrial food system, the fewer of these processed foods I consume. One December, my mom made lunch for me and my dad with pre-washed organic baby spinach that had come from California in a big plastic box. The three of us were sickened seriously enough for me to report the incident to the Canadian Food Inspection Agency after realizing the only food we had eaten in common was those spinach leaves. I learned that these bags and boxes of pre-washed greens are considered a risky food from a food safety perspective—especially if you don't wash them yourself despite the package's claim that they are ready to eat. When the investigator visited my house, he explained to me that when the greens are washed and then packed loose in a box or bag, it creates the perfect environment for bacteria to grow. Apparently, the bacteria that make us sick love a nice flat, humid green leaf to call home. That was the last time I chose to eat those kind of pre-washed greens.

On top of that, these long-distance processed foods simply don't taste good. How often have I picked rotten pieces of lettuce out of my salad in a restaurant that serves these industrially produced baby green mixes? I used to rely on the frozen bags of Thai vegetable medley when I had a baby and was too tired to chop my own red pepper, broccoli, and carrots, but even in my sleep-deprived stupor I knew that what I was drowning in green curry sauce from a tin wasn't very good. The processors seem to freeze only the dregs of the vegetable patch: the broccoli is chewy, the red pepper slices slimy, and the bits of rubbery baby corn that are the highlight of the bag are few and far between. My experience tells me that I don't have to let sanctimony guide me around the supermarket, because if I listen to my taste buds, I end up choosing good foods anyway. In journalist Michael Pollan's bestselling book *Food Rules,* he counsels not to eat any food that your grandmother wouldn't recognize.

We are at a turning point—or perhaps we've made the turn already. We've broken from the historical narrative of food preparation and have outsourced cooking to food services corporations. Less than half of Americans cook one or more meals a day at home, and the numbers are declining. According to the Residential Energy Consumption Survey, only 32 percent of Americans prepared two or more meals at home every day in 2001, down from just under 36 percent less than ten years earlier.[41] And if we don't cook, we are likely buying foods made by someone else. Between the 1970s and the present, the amount of money Americans spend on food outside the home has almost doubled. What this means is that our diet today looks nothing like the food that humans have eaten over the millennia since we made the transition from hunter-gatherers to farmers. Not that food and cuisine have ever been static, but what North Americans eat today, overly processed and produced away from the home, bears little resemblance to the foods humans have toiled over in our millennial kitchens.

This transformation is not just taking place in the supermarkets of the West. In China and India the economy is booming, and the new middle class is tossing out their old food habits and using their higher salaries to pay for a Western diet. The fast-food offerings of KFC are so popular that the Chinese branch of KFC's parent company, Yum Brands, claims on its website that it opens more than one KFC location in the country every day. When I visited Beijing, almost every subway car I rode on was covered with the company's ads, along with video screens playing, in a continuous loop, a commercial that featured happy families enjoying fried chicken. It's a similar story in India. Fast-food joints such as McDonald's and Pizza Hut ring the upscale Connaught Circus in Delhi. According to research from the Center for the Advanced Study of India, at the University of Pennsylvania, traditional staples in the country are being replaced by Western food staples. For example, the chapati, a flatbread made from durum wheat, is often being overlooked in favour of Western-style bread products.[43] In other parts of India, such as in Maharashtra, where sorghum or millet flatbreads have been the custom, the wheat chapatis are taking over. In Aurangabad, at the supermarket in the basement of the city's first mall, which had only just opened when I visited Chandrakalabai, they sold pre-made egg curry, butter chicken, and biryani for takeaway, something people told me wasn't imaginable in the city even a few years earlier. And the locals I spoke with were aghast when I told them that at the food court, the American sandwich company Subway offered a ham sandwich. In that Hindu and Muslim town, no one ate ham—or at least they didn't before. Now Indians, just like the Chinese, are eating more meat, the result of this dietary switch being that production of meat rose 20 percent globally in the first ten years of this century and has tripled since the 1970s, outstripping population growth.

More proof that the diversity of what people eat in countries around the world is shrinking was released by Oxfam in 2011. The organization conducted a global survey, polling sixteen thousand subjects in seventeen countries, and found evidence of rapidly shifting food habits. While traditional dishes such as biryani in Pakistan and paella in Spain made people's lists of favourite foods, pasta reigned as the world's number one preferred dish. Pizza was also popular, and these two meals were among the preferred three foods in just over half of the countries. The report concluded that tastes are changing so quickly that many people are no longer eating the same foods they enjoyed as few as two years earlier.[44]

It's not just a rise in family income that encourages people to drop tradition and drift towards a homogenous global cuisine. There can be an element of wanting to become more modern. When Paul Healy, an archaeologist and professor of anthropology at Trent University, was on a dig of Maya ruins in Belize, the team found thousands and thousands of snail shells amid ancient food refuse. A local man who was working on the dig confirmed that these snails were the remains of a food called *jute* and that today, a few thousand years later, his mother still ate them. Thus began an ethnographic study in nearby villages where Healy asked people about their snail consumption and found that eating *jute* was directly related to age. "Those individuals between twenty and thirty-nine, they never ate the *jute.* But those people over sixty years often ate *jute* and indicated that if their younger family members would get them some, they'd eat them more often. You could see that the young people were even a bit embarrassed that anyone would eat them," he told me. "What it appeared was that collecting and eating *jute* might be a very long tradition—because we were finding them in archaeological deposits dating back to at least 800 BC—and that today it was a practice that looked like it was dying out."

For Healy, the cause of the change was obvious. He has been visiting Belize for research for thirty years and over the course of his career has witnessed the rise of tourism and an associated shift towards a more Western lifestyle. "All life has become more modern and Western. And with that has come a lot of changes in attitude about what is good and what is cool and what is appreciated in a modern society. The food is part of this whole process. When I asked people about it, the young people said, we don't eat *jute* but we eat more modern things, like Spam and canned meats. That was considered better."

The health consequences of these dietary changes have been documented extensively. On the one hand, when people have more money to spend on food, they are more likely to have a more varied diet. However, the Western diet that people around the world are adopting is energy dense and associated with higher rates of heart disease, type 2 diabetes, and some cancers. If these trends continue, by 2040 we'll be diabetic, obese—and we won't know how to boil an egg.

Harry Balzer is an expert on what we eat. Balzer has been studying the American diet for more than three decades at the NPD Group, an American consumer market research firm whose client list boasts some of the biggest food brands. One way the company learns about American food habits is by asking people to fill out a food diary for two weeks at a time. It also goes online to ask 2400 people every day whether they have eaten outside the home. The information it collects is proprietary, and big companies pay top dollar to get a candid look inside the fridges of the nation. So Balzer knows where we've come from over the last thirty years and feels comfortable making educated guesses about where we might be headed over the next thirty. "There are some basic human traits that define how you are going to eat. They are pretty universal," he told me. "How we eat is driven by what we can produce and what the culture knows. You are introduced to that at a very young age and it's very hard to change. So if you have sliced bread in your house, the question is not whether it will be sliced bread that you eat at twenty years old or thirty years old, but who will be the manufacturer of that sliced bread." Even though it's tough to change food habits, there is a good reason why the rest of the world is adopting more and more the Western diet.

When Balzer was hired by NPD Group in 1978, he began by studying what people order when they go to restaurants. Back then, the top three items people chose off the menu were soft drinks, french fries, and hamburgers, in that order. "Jump forward thirty-two years and all we've changed is the order." The top three foods people choose today are soft drinks, hamburgers, and french fries. "Let's jump thirty more years into the future and see what they will be. I'm sure the top two will be soft drinks and hamburgers," he said, giving himself some wiggle room. "Fries could drop to number four." These are the kinds of industrial foods that people around the world are adopting—the quintessential Western diet of pop, fries, and a hamburger. What do they have in common? They are cheap and convenient.

"What will change what we eat is if we can find something easier or something cheaper." Price and the time required to prepare food push change. There are a few other factors that affect what we eat, Balzer added—we humans like to be what he calls "explorers," so we enjoy trying new foods, which explains why the Western supermarket now sells sushi alongside those rotisserie chickens. But to profoundly change our habits, something needs to be easier or cheaper. "The driving force in our lives is time and money."

The origins of this shift in the West to the convenient, low-priced fare of the industrial system go back to that time after the Second World War when to eat packaged foods was to be modern; they were even a status symbol. Over the next few decades, more and more women sought paid jobs outside of the home. In Canada in 1972, 42 percent of women worked outside the home. Over the next three decades the number of employed women in this country rose steadily. With more women earning money, the average household had greater disposable income, and since the women spent their days away from the kitchen, they had less time for preparing dinner. More money and less time made the convenience foods all the more appealing. Why bother making something from scratch?

As well, cooking is seen as women's work and is not valued in our society, along with all that other domestic labour that women have long done with little recognition or appreciation. So why would a woman who now has control of some of her own money choose to spend her precious time in the kitchen doing something that society doesn't consider to be worth much? All the time my grandmother spent in the kitchen as a homemaker was no longer possible for my mother, who worked full-time. So while my grandmother had made vegetable soup from scratch and baked oatcakes and shortbread, in my childhood home the cupboard was always stocked with Campbell's Vegetable Soup (I loved it!) and the homemade cookies I remember most from home either were made only at Christmas or were the ones I started to bake when I was about ten. While my mother did make dinner every night, she didn't have the time or the inclination to undertake the time-consuming food preparation that her mother had done before her because she was working all day at the office and then doing the domestic shift when she returned home.

These are the kinds of family narratives that help Ian Mosby, a post-doctoral fellow in the History Department at the University of Guelph, piece together the logic behind this cultural shift to industrial food. He studies these changes in eating in postwar North America, trying to better understand how they came about. He also collects those community cookbooks that church groups and women's associations used to put together with their favourite recipes to glean a clear picture of what people actually ate. "Preparing food is time-consuming. That is why people experiment with the instant foods in the late 1950s," he said. "In trying to understand why someone would make a casserole with a couple of cans of soup and ground beef—well, it's pretty easy, whether it tastes good or not." With women working, the economy chugging along, and with expanding industrial production processes bringing the cost of food down, the convenience foods were increasingly alluring. "For a lot of people, this is the first time they could afford middle-class food," he said. "Canned soup was too expensive for most people until the 1940s and '50s."

Attitudes were evolving too. "Changing ideas about gender equality has an impact on how long women want to spend cooking," Mosby said. This trend has held over the decades and is borne out in the statistics. A study published in 1991 in the *American Journal of Agricultural Economics* looked at the connections between women's employment and family food expenditures. It found that when women bring their earnings home, the proportion of family income spent on food goes up. The study also found that families were more likely to eat away from home if the mother held a job outside the home. "The rise in the fast-food industry is the same thing," said Mosby. "Women wanted a break."

So in American society people were able to afford to pay more for their food and families started visiting restaurants more frequently. Until not long ago, people in the business thought restaurants were the future for food. "Conventional wisdom in the 1970s was that there wouldn't be supermarkets in 2011," said Balzer. The belief was that, since women were working, people would just go out to eat instead of cooking for themselves. The rise in restaurant visits supported this theory. In fact, business was booming to the point that food companies started to buy restaurant chains. Pillsbury, which was producing products such as cake mixes, acquired Burger King in 1967, and General Mills acquired Red Lobster in 1970. Over the next decade, Pepsi bought chains such as Taco Bell and Pizza Hut. However, this restaurant boom peaked in 2000. The number of restaurant visits in the United States began to drop and has been in decline ever since, said Balzer. But the change in numbers didn't indicate a return to the kitchen, he said. Rather, the take-home business captured the market. "It's a hassle to take everyone out to eat," he said. "It's Isaac Newton's theory 'What is at rest stays at rest' applied to food consumption. A body at home stays at home." This is the first clue to understanding why supermarkets started to look like take-out joints.

In the mid- to late 1990s, the concept of the home-meal replacement was born: purchased home-style meals for people to eat in the comfort of their own kitchens. The idea was quickly adopted by the entire supermarket industry. Soon rotisserie chickens appeared by the deli, jars of prepared soups lined the cooler sections, and packages of pre-cut fruit were sold with the promise of making our life simpler. By 2010, the number of meals that Americans took home almost doubled. "The supermarket makes life easier because the food is prepared, but it's cheaper than a restaurant." Again, those two factors that Balzer says change people's food habits: easier and cheaper.

What this means for the future, according to Balzer, is that people will continue to eat the same foods—unless something easier and cheaper comes along. The number one dinner entree today in North America is a sandwich. "In the future, I am going to be hard pressed to say that a sandwich isn't going to be the number one or two item," he said. But what he can't predict is how that sandwich will get made. "Who is going to be the food preparer?"

If the big question about the future of food is who will prepare it, then the big question for a sustainable food system is: What needs to be done to ensure that the way our food is prepared is sustainable? To prepare food is an act of creation, an expression of culture. So if a globalized food culture in an industrial food system produces industrial food, then what kind of food culture do we need to support a sustainable food system? Fast food, home-meal replacements, and frozen pizza belong to the industrial food model. What do sustainable meals look like? To achieve sustainable food systems by 2050, we must nurture a culture of food that supports the future we want.

The concept of food culture is possibly best expressed by the word *terroir.* This word is most commonly used in the context of wine to describe external factors, such as soil, climate, and geography, that influence the way a grape grows so that a wine has a particular terroir, an almost intangible essence that defines its flavour and body and makes it what it is. Other foods have terroir too—cheese, chocolate, and even meats like chicken have flavours and qualities that reflect the place where they were grown, produced, or raised. They have a taste of place.

But the concept of terroir can also more broadly encapsulate various aspects of food culture. In 2005, UNESCO, the United Nations cultural body, adopted its own, broader, definition: "A Terroir is a geographical limited area where a human community generates and accumulates along its history a set of cultural distinctive features, knowledges and practices based on a system of interactions between biophysical and human factors." This definition recognizes the profound connection people have over generations to the place where they live and the traditional knowledge and customs that are passed on through time, such as agricultural practices, styles of cuisine, and ways of preparing food. Each terroir is different from the next: the mountains of the Aubrac are home to a particular *paysan* agriculture, whereas the mountains in Yunnan or the dry fields of Maharashtra have encouraged the creation of an entirely different way of life. The UNESCO definition acknowledges the value of this cultural diversity that arises out of place as well as the agricultural and natural biodiversity that typically accompanies it. But terroir is not a static concept. The UNESCO definition goes on to say: "The terroirs are living and innovating spaces that can not be reduced only to tradition." Just like nature, which is in a constant state of evolution, terroirs are alive and can change and adapt. Terroir is living, breathing, and changing food culture.

Fundamental to this broader concept of food culture, and possibly most relevant to the building of a sustainable food system today, is the notion that an intimate connection is made between humans and nature when food is produced and then eaten. Amy B. Trubek, an assistant professor at the University of Vermont who has written extensively about the history of French cuisine and terroir, underlines in her book *The Taste of Place* that the significance of eating and drinking goes beyond simply feeding hunger and thirst. "Whenever we eat food, it's when we literally ingest the natural world. We bring the natural world into our bodies," Trubek explained to me. "It is a universal relationship, because all humans must interact with the natural world to feed themselves. The concept of terroir is saying there's something celebratory in the landscape. It says our relationship is about how the soil, the slopes, the rocks, gives us something we taste." It also reminds us that human societies are in fact a part of nature, reliant on the natural world to survive.

The future of food is dependent on who will make it, and the future sustainability of our food system is dependent on our future food culture. By engaging with terroir, we can start to acknowledge the connection between nature and humans that exists every time we eat. We can start to revitalize that "system of interactions between biophysical and human factors." The industrial food system has broken our connection with nature. But once we bring that connection back, we can start to see our place in the natural world and strive for balance. Just like it is already unfolding in France, way up in the mountains of the Aubrac.

CHAPTER TWELVE

The Terroirists to the Rescue! The Pope of Aligot and the French Culinary Resistance

On a sunny spring day, on the outskirts of the French town of Laguiole, a man stood beside a big truck with a shiny metal tank, waiting to empty its load. Inside the tank were thousands of litres of fresh milk that the man had picked up that morning from more than twenty farms that belong to the Coopérative Jeune Montagne. The man attached a hose to the truck's tank to connect it to the large metal reservoirs positioned downhill from where he'd parked; gravity would carry the milk first to the tanks and then into the building where the cheese-making process would begin.

Every day of the year, the co-operative turns the Aubrac milk into the hard cheese they call Laguiole. It makes 800 tonnes of the stuff, following, more or less, the traditional methods developed by the monks and later used by the old cheese makers in the *burons* on the mountains. This cheese is eaten around the world. At one point, it was even sold at Whole Foods in the United States. But one could argue that the cheese is most cherished by those who make it, because this cheese is the centre of the story of how the Aubrac was able to balance tradition and modernity, drawing on its culture of food to build a sustainable food system. The co-operative is the foundation for the cheese. And the man behind the co-operative is Monsieur André Valadier.

Monsieur Valadier seems to always be smiling, wearing an expression of wonder, as if he were curious about everything around him. For the length of my stay, he bounded from one activity to the next, helping with the morning milking, going out onto the mountain with his sons to check on the cows, attending meetings, giving talks—I could hardly keep up. He certainly didn't act his seventy-eight years. Supposedly Monsieur retired in 2008, when his term as president of the co-operative finished, ending a long career that included a handful of public positions such as mayor of his commune and an elected member of the regional council of the Midi-Pyrénées, where he was in charge of the agriculture and rural economy commission. Nevertheless, he keeps on working. He remains deeply involved in the cheese co-operative in Laguiole that he helped to found, he sits on various boards and committees, attends meetings, and participates in delegations that take him to places as far away as Canada and Morocco to talk about the experience of using culture to boost local economies and help to preserve sustainable farming.

Monsieur Valadier lives in La Terrisse, a small hamlet, a cluster of old stone houses, across the way from Les Clauzels. If you follow the narrow road that winds through the town past the church, you'll come across a statue of the Virgin. This just about marks the spot where Monsieur Valadier lives with his wife, Geneviève, in a house that once belonged to her grandparents. It's a beautiful house that has been renovated recently, so it is comfortable, with a modern bathroom and kitchen, and yet retains the olden days charm, with its thick wooden beams and intimate low ceilings. In the living room, they have preserved the old fireplace, which is so tall I could stand upright inside it. On one side, an old bench is built into the base where one could sit to keep warm in the winter. This preservation of historical details is Madame's influence. She grew up in Paris, the daughter of émigrés from the Aubrac, and met Monsieur when she would visit her grandparents' home during summer vacation. She taught her husband an appreciation for old buildings, for the history that lives in the structure. "My wife helped me to see things differently, to see the beauty in a stone wall, in the architecture," Monsieur tells me.

The house's blending of the past and the present is representative of Monsieur's outlook on the world. In fact, he is particularly adept at blending the past with the future. For example, it seemed incongruous that this old *paysan,* who grew up wearing homemade wooden clogs, chose to slip on plastic shoes before he left the house. His ability to adapt to change works at a deeper level too. One day at the co-operative, a man in a white lab coat came up to Monsieur Valadier and passed him a block of cheese he'd recently created a recipe for, wrapped in paper.

"I have a cheese I want you to taste," he said.

"What's it called?" asked Monsieur.

The man in the lab coat smiled sheepishly and shrugged. "I don't know."

"What about Laguiolais?" Monsieur asked spontaneously, playing on the name of the historic cheese.

"Hey, that's good!" The man beamed.

"We'll see, we'll see," said Monsieur.

"The tourists want a cheese they can take with them," he explained to me in his minivan. The traditional blocks of Laguiole are too big for one customer to buy and so they must be sliced to be sold in smaller amounts. But the cut cheese isn't conducive to travel without refrigeration. While a new cheese wouldn't be part of the historical narrative of the Aubrac, it would help the co-operative adapt to the present and make some money while keeping it linked to the past. And as Monsieur would say, as long as they respect the past, they can move forward.

"I've always remembered my parents' advice if I found myself crossing the mountain plateau in terrible weather," he said. "'If, suddenly, you find yourself stuck and can't move forward, all you need to do is turn around and go back. Retrace your footsteps in the snow before they disappear. If you follow your footsteps, you will find your path.' We can do this now too, with our way of life, our agriculture. And we don't have to go back to wearing our old wooden clogs."

Cheese making in the Aubrac has a long history. It began sometime during the Middle Ages, when the Catholic Church was expanding its territory in Europe and monks moved into the area, building their monasteries and clear-cutting the old-growth forests so they could plant their crops and pasture their animals. "It was the monks who came and transitioned our ancestors from hunter-gatherers to farmers," explained Monsieur Valadier. The monks also brought with them Catholicism, and the Aubrac—just like the rest of France—quickly became a land of Christian believers. There is a story people tell of Adalard, the viscount of Flanders, who around 1120 went on a pilgrimage, walking the old pilgrim route to the Camino de Santiago de Compostela in Spain. Back in the twelfth century, the section of trail that crossed the Aubrac was dangerous, because of the snow in the winter and fog in the summer that made it difficult to pass, and because of the bandits who operated there. The story goes that the viscount ran into trouble on his pilgrimage and pledged to God that if he did make it to safety, he would build a monastery hospital in gratitude. He survived, kept his promise, and founded his holy institution in what would become the town of Aubrac.

Over the next decades and then centuries, a new agricultural and economic system developed in the area. The monks who lived in the abbeys in the valley started to send their cattle up the mountain to graze on the grasslands. This way of life was soon adopted by the people who lived there, and they sustained themselves with the cattle that migrated every year from the valley to the summer mountain pastures. Within the next hundred years, it has been recorded that as many as eight thousand animals were taken on the annual transhumance, and by the fourteenth century, the Aubrac had become completely reliant on a pasture-based economy. The monks used the cheese for their own consumption but also for their religious works; it is said that hundreds of pilgrims as well as beggars arrived at their doors every day and were fed. When the French Revolution radically transformed the country and, beginning in 1789, threw out not only the nobility but the oppressive abbey system that allowed the monks to rule the masses with impunity, the pastoral way of life the monks had introduced to the Aubrac nevertheless persisted. Those who lived in the area kept up the tradition of the transhumance and the cheese making. In 1883 in the Aubrac, there were three hundred stone *burons* operating. In these stone huts, more than one thousand *buronniers* spent their summer months producing seven hundred tonnes of Laguiole. Around the time Monsieur Valadier was born, in the 1930s, cheese making was so central to the area that these men formed a union to help with the sale of their cheese to other parts of the country.

During the Second World War, the people of the Aubrac felt the impact of the conflict, despite living in such a remote area. When it appeared that the Germans would be landing on the beaches of the Mediterranean to invade the country, Monsieur Valadier remembers that the French president broadcast a plea over the radio to the people in the south, asking them to blow up bridges to make it difficult for the Nazi soldiers to make their way north to Paris. Later, during the German occupation of the country, the area became a hiding place for those involved in the resistance, and Monsieur remembers an entire village that was burned by the Nazis, who were trying to beat down local support for the rebels.

It wasn't until after the war that Monsieur Valadier left the farm. Up until 1954, he had been overseeing the farming, ever since his father was killed in a logging accident when Monsieur was nineteen. So when he was conscripted into the French army in 1954, he left the hills where he was raised and went to work as a military police officer in Toulouse. He spent the next two years in service as a member of the army motorcycle corps (avoiding being sent to Algeria, where France was fighting that country's independence movement, after he injured his knees in a serious motorcycle accident). In Toulouse, he collected a military salary that was more money than he had ever known. In fact, so much money that, when his two years of compulsory service finished, he returned to his hometown and bought a brand-new tractor. This made a significant impression on the young man. He was appalled that as a member of the military—a job he didn't value much—he was given an ample, steady wage. But as a farmer producing the food people require to survive, he earned little. "It's not right," he remembered thinking. "The role I performed on my farm is more important than the role I was paid to do."

Around the same time Monsieur Valadier bought his tractor, other farmers in the area were changing their practices too. For one, the system of the *burons* was quickly disappearing. The rustic conditions on the mountains were not only uncomfortable but dangerous. "There were accidents, like when a cow butted a *buronnier* with its long horns and the man died because he couldn't get medical attention up in the mountains. People couldn't accept that," said Monsieur Valadier. During the 1950s, the number of *burons* dwindled to as few as fifty or so, down from the several hundred there had been a few generations before. Each year they produced just twenty-five tonnes of cheese as opposed to the seven hundred that had previously been the average. The rural exodus that had begun decades earlier, when the first peasants began leaving on foot for a better life in Paris, gained speed as young people left the Aubrac in search of work elsewhere. Those who stayed behind continued to struggle to make a living. "We said, how will we live?" Monsieur Valadier recalled. They tried to sell the cream off their milk, but they were paid the same low price for what their alpine cows produced in smaller volumes as farmers raising cows in industrial operations at lower altitudes; the terroir of their cows' cream wasn't factored into the price. To make a living, they would have had to produce twice as much cream. They made very little. "It was a crisis."

A small group of farmers got together and decided that the only way to move forward was to try to bring back the tradition of cheese making that had sustained their area for so long. They needed to reconnect with their culture and with the terroir that was intrinsically linked to who they were. So they founded a co-operative to pool their milk and produce cheese. There was the feeling that if they didn't take action now and try to improve their fate, they would all simply fade away with history. The farmers weren't the only ones who sensed that their pastoral way of life was on the verge of disappearing. Apparently the government of France did too, and a team of agronomists was sent to the area to record how people there lived and farmed. They tried to understand how farmers had survived off the land in the Aubrac since the Middle Ages. The presence of the agronomists gave the farmers pause. "This helped us to understand that we were about to move into a new era," said Monsieur Valadier. "They thought we were going to disappear. But they didn't expect the co-operative."

The men who started the co-operative were all young. Monsieur Valadier was only twenty-five at the time and his colleagues were his contemporaries. It was his mother who, when thinking of the group of renegade young farmers, named the co-operative Jeune Montagne, which translates as "Young Mountain." The men transformed a barn that had belonged to Monsieur Valadier's grandparents in the small village of La Terrisse into a rudimentary cheese-making facility and bought some second-hand equipment. Their plan was to make the Laguiole cheese in a similar way to the *buron* method: they would collect the milk from the farmers and turn this raw milk into cheese that they would leave to age. They didn't need to learn how to make cheese because they had all grown up watching their parents and grandparents work the curd. However, they did need money. When the bank refused them a loan, they went door to door in the community, raising the money to start the enterprise. Then in 1960, they made their first cheese.

Immediately, the co-operative started to grow. That first year, thirty farmers contributed their milk to the pool, a number that jumped to 250 over the next five years. Whereas only 25 tonnes of Laguiole had been made in the area before the co-op was formed, in 1965 they produced 100 tonnes of the cheese, in addition to 20 tonnes of the tomme. By 1985, there were 114 farmers producing 590 tonnes; today there are fewer farmers, only 79, but they produce even more cheese.

Local industries spun out of the cheese. The town of Laguiole is also known for its traditional hook-shaped knife that local artisans made for the *buronniers,* the handles carved from the horns of the cows. Tourists come to buy the knives, eat aligot, and of course visit the co-operative. They stay in the numerous hotels and bed and breakfasts and eat in the town's restaurants. One of France's most well known chefs, Michel Bras, is from the region and runs an exclusive restaurant outside of town that has been awarded three stars by the prestigious *Michelin Guide* and is part of the high-end global network of Relais & Châteaux. The restaurant is a modernist building perched on a hilltop overlooking the vast landscape, and Bras is best known for serving exquisite dishes composed of the vegetables from the region, as well as aligot—meals that start at two hundred euros.

But mostly it is the terroir the visitors come to experience. "The tourists' first request is to see the cows roaming free in the fields!" said Monsieur Valadier. "We never expected that."

There is one central reason that the farmers can live by their terroir. Government policy has helped them to realize the value of their geography. Said Monsieur Valadier, "If there wasn't the *appellation d'origine contrôlée*"—or controlled designation of origin—"we wouldn't exist." The town's story demonstrates how governments can use cultural policy to ensure that terroir can help to preserve an integral part of what makes us human and build sustainable food systems.

An *appellation d'origine contrôlée,* or AOC for short, is a French certification that officially recognizes the role of terroir in the creation of a product. (Other countries have similar appellation systems.) Under the system, a type of wine, or a food such as a cheese or a breed of poultry, is legally designated as an official product of a region. This designation stipulates how the food should be produced, with which ingredients, and that traditional methods be followed in a specific place. It is essentially a trademark, or a brand, that allows consumers to differentiate between similar products and helps producers capitalize on the distinct geography and culture of their region. This French certification system is governed by a national institute that falls under the Ministry of Agriculture, Food, and Fisheries. It first drew up these standards and now enforces them from its office in Paris.

The roots of this concept can be traced back to the fifteenth century, when people started to make Roquefort, a creamy blue cheese aged in caves in the eponymous town, not too far from the Aubrac. The people who made Roquefort wanted to ensure that cheese makers in other places didn't try to sell their blue cheeses under the Roquefort name. To protect their trademark cheese and to ensure that others didn't take advantage of their unique brand, the state granted the people of Roquefort a monopoly. Other cheese makers were forbidden to call their blue cheeses Roquefort. It was several centuries later, in 1925, that the state of France created the legal AOC label, and the Institut National des Appellations d'Origine (now called the Institut national de l'origine et de la qualité) designated Roquefort the first AOC-certified food, enshrining its uniqueness in the new law.

There are now forty-five protected cheeses in France, three butters, two creams, more than three hundred wines, and even some animal breeds. Laguiole cheese received its appellation in 1961. To earn an AOC designation, applicants are required to produce what's called a *cahier des charges,* a document that lists production standards with the aim of protecting the original essence of the product. In the case of Laguiole, the *cahier des charges* delineates the geographic area where the cheese can be produced and specifies which breed of alpine cattle can provide the milk and the conditions under which they are to be raised, including what they can be fed. But these standards are not necessarily static. Over the years in the Aubrac, the requirements have been adjusted to reflect changes in farming. For example, while traditionally the milk from the Aubrac breed of cattle was used to make Laguiole, the Aubrac today have lost their ability to be milked and are beef cattle. So according to the *cahier des charges,* milk from a similar heritage breed of alpine dairy cattle, the Simmental, may be used to produce Laguiole.

If Roquefort needed trademarking way back in the fifteenth century, in today's worldwide economy of food, the protection afforded by the AOC plays an even more important role. "With globalization, we risk losing everything," said Catherine de Menthière, a former associate director of the institute that oversees the AOCs. By stipulating the ingredients and steps that must be used to produce one of these certified foods, you shelter the recipe from passing trends. "We protect it from the fashion of the day. If you put a little salt and a little sugar and a little colouring, with time you forget the original recipe," she explained. "We consider these recipes to be our culinary heritage, equivalent to our churches and our castles. The AOCs are part of our national heritage."

The regulations laid out in the *cahier des charges* are not a homogenizing force, and there may be variations in a product. "Depending on the seasons, the cheese has different tastes and colours," explained Monsieur Valadier. "A cheese that has been made in the summer, from milk of the cows who graze in the grass, this cheese has a deeper yellow colour. The texture is a little more subtle and the taste has more character, with notes of vegetation. A winter cheese is more white with flavours that aren't quite as complex." That's the beauty of cheese made by hand: each batch is a reflection of its environment and circumstance. The flavour is affected by the milk's fat and protein content, by the season, by the way the cheese maker works the curd that day. The result is that the AOC rules exist not to create a uniform product but rather to preserve the traditions, the culture, that give rise to something organic, something made by hand, something that reflects the changing nature of terroir and, through it, our connection to nature.

The AOC certification process is a cultural policy. It's an example of the way a government can foster consumer support for a particular kind of food system that can't compete with industrial food—not because it is not efficient but because the way of buying and selling food in the mainstream has no way of attributing value to the ephemeral qualities of terroir. AOC certification helps farmers to charge a price that reflects the different way they produce their food. "The government wants to help farmers to make a living by providing value to farming," said de Menthière. "It permits them to localize a product, because this product can only be made in that geographical place." Without the AOC trademark, as Monsieur Valadier pointed out, the farmers of Laguiole couldn't compete with industrial cheeses. When a company produces cheese in a factory, its goal is to produce the best product it can at the lowest cost—the cost being a financial tally of the price of the various ingredients plus labour, equipment, et cetera. However, Laguiole is produced in another paradigm entirely, one that is based on culture. The farmers aren't focused on using the least expensive inputs; rather, their priorities are a way of life that is in tune with the cycles—and limits—of nature. In the Aubrac, the cows that thrive best in the harsh conditions of the mountains typically don't produce as much milk as, say, a Holstein cow that does well in a different environment. An economist might say that the Aubrac cows aren't as efficient at producing milk and the people of the region should be buying their cheese from some other place that has a comparative advantage, a geography and climate that is conducive to cheese making in abundance. But Laguiole has a value that is more profound than simply dollars and cents.

There is a French concept in the field of microeconomics that captures the notion that a product can have an intangible value that consumers are willing to pay for—*un panier de biens,* literally "a basket of goods." When Monsieur Valadier says that the value of his region's milk is more than what it's worth on the open market, that it has *un panier de biens,* he means that people are willing to pay extra for something they can't necessarily touch or taste in the cheese itself.

In the *panier de biens* model, there is a direct connection between the quality of the cheese and the environment. To make the product that customers want requires a particular care for the land and environmental stewardship. In economic speak, this externality can be internalized into the price of that cheese, allowing farmers to charge more and, in the process, earn more money for being good environmental actors.[45] The farmers—by practising traditional low-impact grazing that allows the local biodiversity to flourish, by preserving the gene stock of the small cattle breeds, by attempting to live in sync with the natural systems in the area, and by preserving the landscape—are protecting the environment of the region and offering society ecological goods and services. Said Monsieur Valadier, "The terroir-based approach always takes the natural environment into account. The soil, the climate, the flora, the genetics. It necessarily plays the role of actor in the production of the cheese." The AOC puts an official stamp on this elusive process, and that allows the farmers to make a living. It supports an alternative and regenerative way of farming.

The benefits have been repeated many times in other parts of France too. A recent addition to the list of AOC cheeses is a soft goat's cheese called cabécou de Rocamadour. It is produced in the vallée du Lot, an area that at the end of the 1980s was experiencing a decline. Only about forty producers kept small herds, producing cheese for the local market. The cabécou de Rocamadour received its AOC certification in 1996, and within only two years, circumstances had changed. There are now 120 producers on eighty farms making cheese with the milk from herds that had swelled in number to ten thousand goats.

At the same time, AOC designation doesn't shut the country's large dairy corporations out of the AOC cheese market. Anyone can make an AOC cheese. You don't need to be a small-scale artisan to receive the official stamp. It's just that everybody must follow the same guidelines when making their product. And these guidelines are written to favour a certain kind of production.

The widely known camembert is a soft cheese, a creamy round covered with a supple white mould that is developed by the introduction of the bacteria *Penicillium candidum.* This style of cheese was created in 1791 by a farm woman and has since become one of France's most famous, both internationally as well as at home. According to the *Oxford Companion to Food,* the cheese secured its place in national myth when it was distributed to French soldiers during the First World War. The term "camembert" is a generic one and it belongs to the public domain—this was confirmed by the French courts in 1926 after the camembert makers lost their case for an exclusive trademark for the name. This means anyone can make a cheese and call it a camembert, and many industrial cheese makers do. However, a Camembert *de Normandie* (emphasis is mine) is an AOC-protected cheese, and to produce a cheese with this name, you must follow the procedure set out in the *cahier des charges.* You must begin with raw milk that has a fat content of at least 45 percent and must make the cheese by hand, as well as use a special ladle to pour the salted curd into a mould in three stages before it is set to age. The world's largest dairy producer, the French Lactalis, was interested in making this cheese and lobbied for the rules to be changed so it could make a pasteurized version. In 2008, it lost. Today, the company instead makes its Président brand of pasteurized camembert, which is not AOC certified and is sold at a much lower price than what you pay for AOC-certified Camembert de Normandie. This is an example of how the AOC system allows farmers and artisanal producers to derive value from their distinct geographic and cultural circumstances and then compete as economic actors in the marketplace, even against large corporations that are given equal opportunity as long as they follow the rules.

The success of the AOC system couldn't exist without a buying and eating public that supports the idea that it matters where your food comes from and how it is produced—beliefs that are of fundamental importance in France. While some people in the country do worry about whether France's gastronomic heritage can weather globalization, to those of us on the outside, there are still plenty of signs that it has a healthy future.

Rodez is a small city an hour and a half's drive from Monsieur Valadier's La Terrisse, built around a sixteenth-century cathedral. In the old centre of town, the streets are too narrow for cars, and today people meander down the old medieval roads that are lined with many food shops. When I visited, it was just before dinnertime in the late spring, and everywhere there were indications of the importance people in the town put on the locality of their food. The sign above the cherries at one stall indicated in which region of France they were grown as well as their variety. The poultry in the butcher's glass case was labelled by breed, and even in the supermarket, the Comté cheese carried the AOC label. There were also shops offering the *goût du terroir* to tourists, many of whom would likely have been French travellers. The French are big tourists in their own country, and in fact, according to Amy B. Trubek, the professor at the University of Vermont who studies the French relationship to terroir, it was culinary tourism in the age of the automobile that helped to develop and perpetuate the idea of regional gastronomy.[46]

One of the last century's great writers and gastronomes was a Frenchman who went by the name Curnonsky. (He makes an appearance in Julia Child's autobiography as the elderly writer wearing a billowing nightshirt and eating a boiled egg whom she meets at his apartment with her co-authors.) Curnonsky wrote about the different specialties in France's various regions, which it turns out, according to Trubek's book, was in part to encourage travel within

France. To this day, masses of French tourists explore all corners of the country, staying in rental holiday accommodation, *gîtes,* and bed and breakfasts, *chambres d'hôtes,* with their Michelin Red Guide in hand to help them find the right place to eat. The iconic Michelin guide was created by the tire company in the 1950s to give people a reason to get out and see the country—and meanwhile burn some rubber. What better reason to go for a drive than to discover the taste of place in another region, a taste so distinct it can't be separated from the geography. While travelling in France, to this day you will still frequently hear the expression "*les spécialités de la région.*" And because of the long-standing French tradition of small, artisanal production, there were all sorts of delicious foods to discover on the road, and still are today.

All this demonstrates that food culture is something that needs to be taught, cultivated, and supported. And if the French can do it, so can others.

Around the world this same type of connection to the earth roots people of all backgrounds and cultures to their corner of the planet. However, an industrial food system that sees peasant production as inefficient and land as something to be developed has threatened this connection and in many cases ruptured it entirely. Parviz Koohafkan recently articulated a concept while working for the Food and Agriculture Organization that he hopes will help to motivate governments and institutions, as well as the broader public, to value the places in the world where this connection is most salient and preserve them.

Originally from Iran, Koohafkan received his PhD in terrestrial ecology from the University of Montpellier in France and was hired by the FAO in 1985 as chief technical adviser in natural resources and watershed management programs in Latin America and the Caribbean. For his work, he frequently visited indigenous communities in Ecuador and Peru who lived in the forests of the Andes, supporting themselves through an intricate management of the ecosystem. While it was a life of poverty for the indigenous farmers, Koohafkan noticed that they were able to produce enough to eat by using their traditional knowledge to draw from the biodiversity of their landscapes. "Some ten years later, when I returned to visit, many communities and their agricultural heritage were no longer there," he told me. The people and their way of life had disappeared, a pattern he saw repeated again and again, in his own country of Iran, in Kenya, where he also used to work, and in India and China. "Unfortunately, the changes do not bring improved livelihood, poverty reduction, or food security. Most of the time, migration meant the loss of many natural goods and services as well as the destruction of the traditional agricultural systems." The ways of farming that people had created over millennia, existing in a state of true sustainability for generations, were disintegrating along with the knowledge of how they worked.

Koohafkan decided that the global community must do something to help preserve these landscapes. He coined the term Globally Important Agricultural Heritage Systems—also referred to as GIAHS (pronounced "ghee-ahs")—to describe the many ingenious ways humans have found to produce food while remaining deeply integrated into the broader ecosystem, such as a forest, a desert, a mountain range, or a shoreline. In 2002, he proposed a partnership between the Food and Agriculture Organization and member countries to focus international attention and resources on preserving these landscapes. Since then, about twenty countries have joined the GIAHS program, and up until his retirement in 2012, Koohafkan was director in Rome, at the FAO.

The program divides heritage agriculture systems into ten categories. These include the intensive spice-cropping systems that require extreme care and finesse, such as saffron production in Kashmir, and polyculture farming systems, in which a number of crops are grown together, the varieties selected for their ability to handle a variable climate. They also included understorey farming, found in places such as the South Pacific island of Vanuatu, where people have managed fruit orchards in the forest, tending to crops and wild foods on the forest floor; as well as the nomadic and semi-nomadic pastoral systems in which communities use pasture, water, salt, and forest to raise their animals, such as yaks in Tibet and Mongolia or reindeer raised by the Sami in Scandinavia's Far North. Also on the list are hunter-gatherers like the honey-collecting forest dwellers of Central and East Africa.

Probably the most salient distinguishing feature of these agricultural landscapes is that that they've lasted millennia. These agroecological landscapes provide food security in both the short term and the long term. Not only can people eat today, but by protecting the earth that yields their food, they have also ensured that their future generations too will have something to eat. And then there are the more intangible benefits of biodiversity. These landscapes are havens of agricultural and natural biodiversity. They are treasure troves of genetic material. They are the essence of sustainability. They show us how it is possible for humans to feed themselves in one place, over a period of thousands of years, without exhausting the natural resources around them. They demonstrate that it is not in our human nature to destroy all that is around us. One of the secrets to the longevity of these landscapes is the intensive role people play in these ecosystems, managing water sources and soil fertility, encouraging biodiversity within agricultural crops, and protecting the wild habitat around them. The knowledge of how to do this is passed down through the generations. It's an important part of their cultures. What these food systems teach us in our search for sustainability by 2050 is that culture is at the root of a sustainable relationship between humans and nature.

They also demonstrate just how, well, awesome and ingenious we humans can be. When I read about what people have dreamt up and then implemented without any help from modern technology or fossil fuels, my spine tingles with excitement. It's unbelievable. Take the ancient qanat water channels built about three thousand years ago first in Iran and then in Afghanistan and other Central Asian countries as well as in Egypt. These underground channels use only gravity to carry water from aquifers in the highlands to arid areas several kilometres away. The irrigation water the tunnels provided allowed for a diverse cropping system, and in some places they also became the home to a breed of endemic blind fish that people could eat too.

Another awe-inspiring irrigation scheme can be found in Sri Lanka, where villages built large water tanks that they connected to their rice paddies, collecting water during the rainy season that they could use after the rains had passed. Then there are the oases of the Maghreb desert in Tunisia and Algeria where Berber farmers have constructed a sophisticated irrigation infrastructure that has permitted people to grow date palms, figs, olives, apples, vegetables, forage, and cereals in the hot sun and sand. Around Lake Titicaca in Peru and Bolivia, you can still find the ancient waru-waru, a system that uses ingenious trenches that are built around raised beds. The trenches fill with water, and because the water retains the heat of the sun, it moderates the temperature in the beds, extending the growing season. Because of the way the farmers have maintained the soil fertility, waru-waru cultivation has been found to have a higher productivity than northern Pampas fields enriched by chemical fertilizers. As we build our sustainable food systems, we can strive to incorporate a similar ecological balance into the ones upon which the traditional landscapes were founded.

At first, these GIAHS sites may appear to be relics of a bygone age of agriculture, sure to disappear in the era of globalization. Certainly, they represent peasant agriculture as it has been practised for millennia. However, they are not inconsequential. Heritage agricultural systems cover an estimated five million hectares, and more than 1.4 billion people rely on them for their food.[47] However, this number is not static, because these peasant farmers, as well as their land, are at risk from land grabs, migrations, and environmental destruction, among other threats. Still, according to Koohafkan, these systems produce a substantial amount of food consumed in the developing world and thus offer an important element of food security. To try to ensure that these areas last into the future, the GIAHS office is helping governments to improve their policies regulating development, agriculture, and land and water tenure. It also works with small farmers directly, creating eco-labelling systems—similar to the AOC system—that enable them to extract the economic value from their unique agricultural situations. And it supports the development of ecotourism. The ultimate goal of the project is to protect all these biological resources while respecting the traditional cultural practices that have maintained them for so long, thus preserving these examples of sustainability.

But it won't be easy to protect this agricultural heritage. People who live in these places are poor. Even though they might have food security because they can take care of most, if not all, of their nutritional needs, by modern standards they are subsistence farmers who often must struggle to have enough to eat. These are also the same people who are leaving their rural communities to join the millions in the factories so their families can pay for things like water heaters and televisions. One reason young people migrate away from their communities is because the way they produce food, as a product of their terroir, is not valued. There is a clash of food cultures.

And we've seen those before.

After Monsieur Valadier bought his tractor in the 1950s, he and his neighbours willingly readied themselves for the transition to modern farming techniques. The farmers of the co-operative heeded the advice of the Chambers of Agriculture as well as the government, who told them in the 1960s that it was time to join the present. "They said, 'You are wasting your time.' We were doing in 1960 what others were doing in 1920," said Monsieur. "As president of the co-operative, I was in charge of diffusing all the ideas from Paris." And his colleagues listened. Whereas people had always raised the Aubrac breed of cattle, outside agronomists told the farmers that there were better, more modern animals to be raised that could provide higher volumes of milk than their old-fashioned breed. They suggested they replace their herds with high-productivity Holstein cattle from the Netherlands, and Monsieur Valadier and his neighbours obediently made the switch. But these high-productivity cattle wouldn't do well grazing on pasture. Rather, the farmers were told they could achieve even greater yields of milk by moving away from pastured livestock and feeding the cows silage.

Silage is made from corn or green fodder crops that are harvested and compacted for storage in airtight silos or large plastic bags (or, more recently, deep pits dug into the earth and lined with cement), to be used to feed livestock during the winter when there is no fresh grass. Unlike hay, which is dried in the fields before it is baled, silage is kept wet so that in the anaerobic environment of the silo (or bag or pit) it ferments. Farmers choose silage because it allows them more control. To hay a field, a farmer must wait for a clear patch of weather, because if the drying crops are rained on, down goes their protein content, and the feed isn't as nutritious. With silage, farmers can work on their own schedule rather than Mother Nature's because they don't have to worry about the weather. So this way of farming was adopted in the Aubrac. In the 1970s and '80s, the seed companies arrived too and sold special varieties of corn. They encouraged the farmers to abandon their pastures completely and instead feed corn or corn silage to the cows. So some farmers started to grow corn in the fields they used to hay for winter fodder.

While the farmers in the co-operative changed their agricultural techniques, they remained faithful to their traditional recipes. Despite pressure from the authorities to pasteurize the milk before producing their cheese, the co-operative continued to make their raw milk cheese with pride. It wasn't a resistance to change—their switch to new farming techniques proved that they were happy to adopt new ways. But they knew they just couldn't make the same delicious product if they boiled the milk first.

Then the trouble began. One day in 1995, the local paper ran a shocking headline: *Bugs in Our Aligot.* "Ooooh," groaned Monsieur Valadier when he recounted the story, stretching out his arms to illustrate the size of the problem. In fact, there was something wrong with the cheese used to make the traditional aligot. There had been a few cases of enterotoxins—bacteria that cause food poisoning—in the tomme, and some people had become sick from it. Not seriously ill, but some had experienced enough digestive problems after eating the cheese to arouse the interest of the media. That same week, some of the co-operative's largest clients, restaurants in Paris, called to say that someone had anonymously mailed them photocopies of the news article. The authorities said the way to fix the problem was to pasteurize the milk. "They said, 'You are finished. You can get rid of your problem only with pasteurization,'" said Monsieur Valadier. Rather than hiding from the facts, the co-operative faced them straight on. Monsieur Valadier called a press conference and announced that they would be investigating the problem.

Over the next weeks and months, the co-operative conducted a series of tests. They learned that on dairy farms where cows eat silage, there is ten times the risk of pathogens entering the milk supply than there is on farms where cows are fed grass and dry fodder. If the milk is pasteurized, pathogens are killed, but Laguiole is a raw milk cheese. So the co-operative analyzed the milk from their silage-fed cows and compared it with that from cows raised the old-fashioned way and confirmed what they'd learned. The problem with the bacteria existed only with the milk from silage-fed cows. The authorities were wrong. The farmers didn't have to turn away from their old raw milk recipe to remedy the situation. As long as they respected the old ways of raising cattle, and stopped feeding the cows silage, they could continue to make their Laguiole just as people had done for centuries.

Soon after the crisis, Monsieur Valadier returned to haying his fields once more and letting his cows out to graze. The co-op rewrote its AOC specifications to forbid the feeding of silage and corn. In so doing, the farmers both reconnected with the traditional knowledge of their ancestors who had first developed these systems and reconnected to the land.

All this could not have been achieved without the support of a terroir-conscious eating public, the people who buy their cheese. And there wouldn't have been an eating public that cared about terroir without the strong support of state policy that prioritizes this culture of food and helps to transmit it. The AOC program helps to formalize the kind of values that thrive in the alternative economic structures. At the same time, they support sustainable farming systems. In our pursuit of sustainable food by 2050, we need to develop our relationship with terroir to strengthen what we are assembling.

CHAPTER THIRTEEN

Culinary Biodiversity: You Are What Your Ancestors Ate

The cows walked calmly towards us. There were eight of them, young and small and curious. When they reached the patch of field where I was standing with a biologist named Mario Duchesne, they stopped about a metre away and leaned their noses towards me, trying to sniff my arm, just like a dog would. "Is this the first time you're seeing them?" Mario asked of this rare breed of cattle, the Canadienne. Mario and I were standing in the wet grass in the field adjacent to his farmhouse in Quebec's Charlevoix region. We wore bright blue plastic bags on our feet to protect our shoes from the muck and stood admiring the eight cows assembling around us that are part of the genetic project to bring back the Canadienne cattle, a breed that, without care, would be on the verge of extinction.

Mario is the managing director of the Association for the Development of the Canadienne Cattle Breed in Charlevoix, an organization that is leading the efforts. The cattle are special because they are the descendants of the first cows brought to North America by French settlers back in the 1600s. Over the centuries, the cows have adapted to the rugged geography and harsh climate of Quebec, and because of this environmental pressure and their isolation from other genetic lines, they became their own breed—one that is connected to the terroir of the area.

In fact, it *was* the first time I'd ever seen a living Canadienne, and they did look different from other cows. Their frames were smaller than the average bovine and their coat was a deep brown, with russet undertones. Their ears were pointed, and they had a gentle, inquisitive look about them. "They're beautiful," I told Mario. He smiled and giggled like a proud father. "That's the original colour," he said. "It's russet and black. It changes depending on the season. We think that is the original colour from before the cow was domesticated." The white or beige hides that are the most prevalent phenotypes in cows today are the result of human selection, he explained. "This one here"—he motioned to a cow that was losing interest in us and turning away—"is three-quarters pure and one-quarter Holstein. Her grandmother was a Canadienne but her grandfather was a Holstein." He pointed to one of the cow's white patches, which indicate her Holstein past. Because there are so few Canadiennes left, Mario would keep this cow despite her mixed ancestry and breed her with a Canadienne bull. He planned to continue to mate future generations of females with pure Canadiennes to slowly erase as much of the Holstein from their family line and bring back the heritage breed. "One day her descendants will be pure."

The desire to preserve the gene stock of rare breeds of livestock and propagate it is about promoting genetic diversity. Ever since humans domesticated the first farm animals, we have developed all sorts of different breeds. But over the past sixty years, we have lost the majority of these gene lines. Once, farmers raised a variety of chicken breeds, and different regions developed their own cattle or goats, sheep or hogs; in Asia, different breeds of fowl can be found all over the continent, suited to the varying climates and geographies. In North America, in the 1800s and early 1900s, entrepreneurial farmers developed many new poultry breeds. One, the Jersey Giant, a big bird resulting from a genetic cross between several types of Asian chicken, was bred in the 1870s in New Jersey to compete with the turkey market. But the industrialization of agriculture included winnowing down genetic lines and promoting one "super" breed over all others. The focus was on breeding animals that would best convert feed into either muscle (meat) or milk; for hens it was muscle or eggs. This was a different approach from the old breeds that provided both.

For meat hens, the super breed was the White Rock, for hogs it was the Yorkshire, and for dairy cows it was the Holstein. Originally bred in the Netherlands, where the fields are flat and where cattle can graze outdoors well into December, this breed has been preferred by agronomists since the 1960s because of the high volume of milk it produces compared with other cattle. Over the next decade, dairy-breeding programs in North America were quick to focus on the Holstein, just like in France, where they gave up the Aubrac and other local breeds. Canada, where the Canadienne has dwindled, is now well known for producing some of the best Holsteins in the world and sells embryos and sperm on the international market.

But when one breed is favoured, the others fade away. In both the Aubrac and the Charlevoix, this departure from heritage breeds left the local gene stock lacking. In the Aubrac, farmers stopped milking their cows and bought Holsteins instead, which meant the old breed became primarily an animal raised for its meat, and it lost its ability to give milk. In Canada, until the 1800s all milking cows were Canadiennes. Then farmers began to import Ayrshire, Guernsey, and Jersey cows from Britain, and later the Holstein, for milk production. These cows were preferred over the Canadiennes, and eventually the old breed slowly began to disappear. Today, breeding pure-line Canadienne cows—or Aubrac cattle—is a rejection of the homogenization of the dairy industry and the worldwide dominance of the

Holstein. It is about promoting pluralism and valuing each breed of cattle for what it offers. "This is part of Canada's heritage," Mario beamed. "There are more than four hundred years of history here. It's exceptional!"

And who has been to visit the Charlevoix to give them guidance and inspiration? None other than Monsieur Valadier.

Mario's farm is down a back road that cuts through bush and farmers' fields, partway between the small town of Saint-Hilarion and the banks of the St. Lawrence River; it's also a short drive from the Laiterie Charlevoix, a dairy that produces a handful of cheeses including one called 1608 made exclusively from the milk of the Canadienne cattle. Jean Labbé is one of the three brothers who run the dairy, having taken over the family business that was started by their grandparents in 1948, pasteurizing and bottling milk as well as making cheese. In 1995, Jean and one of his brothers decided they needed a new plan for one of the last small dairies in the province. Labbé recently had returned from Quebec City, where he had worked for a multinational pulp and paper company for thirty years. He wanted to join his brothers in the Charlevoix, the area he refers to as "*mon pays*"—my homeland—a statement that emphasizes his personal connection to the land. It was soon after his return home that he became interested in the Canadienne cow.

When the brothers took the helm, the Laiterie Charlevoix had been making cheddar since the early days of the dairy, using milk from local farms. But agriculture in the area had changed a lot since his grandparents produced their first batch. The land of the Charlevoix is mountainous, the soil sandy and pebbly and not very productive, suited only to pasture-based agriculture. Until the 1950s, farming here was practised for subsistence. On their small parcels of land, people grew their vegetables as well as fodder for the little livestock they raised. When there was extra milk, they sold it to the dairy to make some money. The dairies then turned the milk into cheddar. After the 1950s, as agriculture was industrializing, many of the farms in the Charlevoix disappeared; those that remained grew bigger, and a couple of dairy farms kept as many as a hundred head of cattle (a number that is small by provincial standards but still large for the area). During the same period, the big food companies were buying the small, family-run bakeries and cheese shops and other dairies. Those who didn't sell mostly ended up going out of business because they couldn't compete with the new commercial products and the arrival of the supermarket. The small Laiterie survived in the face of the trend because it sold its milk in an area no one else wanted to service. Labbé's father drove his milk truck up the north shore of the St. Lawrence River to remote communities where roads were few and a winter storm could see him snowed in, away from home for five days. These risks, though, guaranteed them a market and, inadvertently, a future for their dairy.

When Jean Labbé arrived in 1995, though, the Laiterie's future was uncertain. In the late 1980s, the government had built roads up north, and now all the big dairy companies were shipping their product that way, cutting into the family's sales. More critically, the local farms that had supplied the Laiterie with milk were closing. Labbé and his brother were trying to figure out how to proceed when the two largest dairy farms in the area shut down. "We said to ourselves, if we don't do anything, dairy farming here will disappear," he told me. "We started to think about what we could do." He realized that to find their niche in the market required specializing—just like his father had done. Expanding their cheese production seemed to offer the best promise. But not just any cheese would work. "Because we are small, for me to put a cheese on the market and sell it in the supermarket—it's impossible. I can't compete with Saputo"—one of the largest dairy companies in Canada. "So we have to make ourselves stand out." Most of the cheese in the supermarket, he knew, was made in an industrial setting with milk from the Holstein cow. "We started to look beyond the Holstein," he said.

Enter the Canadienne, a breed of cattle that was a historical product of the region but one that was fast disappearing.

The Charlevoix is a rugged landscape. A thick coat of boreal forest—Jack pines, black spruce, aspen, and white birch—covers the mountains as they slope down into the wide basin that stretches from Jean Labbé's city of Baie-Saint-Paul all the way north to the municipality of La Malbaie along the north shore of the St. Lawrence. It's about a hundred kilometres downriver from Quebec City. The waterway has brought many people here over the centuries. First Nations communities used to catch whales in the river, and in the 1500s, Basque, Norman, and Breton fishers came here too. Later, the European fur traders, *les coureurs de bois,* would have moved through the area on their way to trap beaver. People of the Maliseet First Nation travelled from as far away as New Brunswick to hunt here, and Innu families once spent their summers in the Charlevoix. It wasn't until decades after Quebec City was founded in 1608 that European settlers started to clear the forest and the colonialists began to settle permanently, turning to farming to survive.

It was around this time that the ancestors of today's Canadienne cattle arrived from France. When Samuel de Champlain founded Quebec City, he reportedly kept seventy "*bêtes à cornes*"—horned animals—for their milk. These first cows, however, were likely killed during the siege of Quebec City that began in 1629 and lasted for years as the British and the French fought for the territory. (The

French won.) It was probably in the 1660s when the cows that are the ancestors of the Canadiennes that remain today started to arrive in New France on merchant boats from Brittany and Normandy. Some may have even come on the same boats as the young single women known as *les filles du roi,* the king's daughters, that Louis XIV sent over to marry the predominantly male French settlers in New France. The ancestors of today's Canadiennes originated in northern France where the boats departed from, and some believe the animals could even be descendants of the cattle the Vikings took to France in the tenth century. "It's a hypothesis," said Mario. "But we know that all along the Channel, the Vikings settled and brought their livestock with them. Today, it's the Canadienne colour that is identical to what you see in Norway and Sweden. It's incredible. Is it because they come from the same stock? We don't know. But the theory goes that there is a phylogenetic link back to the Vikings." Not only is this potential link to the Vikings of historical interest, but it could be significant in terms of biodiversity too. "All their relatives have become extinct in France," explained Mario. The Canadienne gene stock is all that is left.

Historians don't believe that too many cows were brought from Europe by ship—maybe only two hundred animals. It would have been difficult: people had to carry enough water and forage to keep the cows alive for a voyage lasting at least three weeks. It would also have taken a few decades to import enough cows to New France to maintain a large enough population in the colony. This meant that early in their time in North America, the European cows would have been isolated from other bovine gene pools. The long, cold winters and short summers of the northern region ensured that only the strongest animals survived to pass on their genes to the next generation. Climate exerted a strong evolutionary pressure on this isolated group of cows, forcing the population to adapt quickly to their new environment. In this way, a new breed was created. And as they adapted to the terroir of Quebec, the cows became robust and reliable and well suited to the rugged life in New France. They were extremely tolerant of the cold and were even able to survive long winters when there was little food. They also became known for their high fertility rate and for being able to birth a calf without any help, a quality prized by farmers to this day. They tended to live long. It was not unusual for a cow to reach fifteen years of age, and some even lived more than twenty years.

According to Mario's association, the Canadienne cows didn't have any exposure to other breeds until the 1850s and '60s, when people once again started to import cattle from Britain. The Ayrshire, Guernsey, and Jersey breeds were preferred by the English-speaking community who largely abandoned the old lines on their farms near Montreal and Quebec City. It was the French-speaking Québécois, particularly the people who lived in remote areas, who continued to raise the Canadiennes. But even these cows started to diminish in number when farmers bred them with Holsteins. In 1850 there were three hundred thousand Canadienne cows in North America. By 1970, somewhere between five thousand and ten thousand cows remained. The last herd of Canadiennes was sold in 1982. The cows I met on Mario's farm are among the five hundred Canadiennes left in Quebec today. Despite their numbers, the hope is that one day a healthy population of Canadiennes will once more be producing milk in the Charlevoix.

A few kilometres from Mario's farm, down the gravel road past fields and forests where there used to be other small farms, is a dairy barn filled with a living, breathing, milking herd of Canadiennes. The herd belongs to Steve and Mélissa Tremblay, dairy farmers who forwent the Holsteins they had been raising for nine years to join the association in 2007 and raise the Canadienne; they sell their milk to the Laiterie Charlevoix for cheese making. It wasn't an easy decision to switch to raising the Canadiennes—Steve's colleagues in the dairy business mocked him for choosing a breed on the verge of extinction over a sure bet, the Holstein. But adopting a new breed offered the promise of participating in something new and exciting—not to mention that the milk of the Canadienne is worth more. The Laiterie Charlevoix pays fifteen cents more per litre for the milk of the purebred. And because the Canadienne milk has a far higher milk fat content, farmers are paid an additional ten cents a litre for the rich milk. This works out to the Canadienne milk being worth twenty-five cents more a litre than the conventional stuff.

Today, the Tremblays' forty head of cattle produce six hundred litres of milk per day that the Laiterie Charlevoix relies on to make the cheese it named 1608, after the year Quebec City was founded. It has recently started to make a raw milk cheddar from the Canadienne milk as well as a soft cheese. When Mario and I walked into the barn the afternoon I visited, the cows craned their necks to see us. The radio played as they chewed their cud. "Steve says they really like it!" laughed Mario. The animal nearest to me stretched out as if to ask for a pat. I stroked her nose. "They are very rustic animals," said Mario. "They have personality. They are people friendly."

The secret to producing more Canadienne milking herds like this one lies with the bulls. A cow can give birth once a year, whereas a bull, like most males, can father many offspring at the same time with hundreds, if not thousands, of cows with no physical risk to its own health. Breeders have exploited this quality—possibly too much. In the 1980s and '90s, a famous Canadian Holstein bull named Starbuck was so fine—he was large in size and tended to father female offspring that produced high volumes of milk—that the sale of his semen to breeders around the world netted a whopping $25 million. Starbuck was aptly named because he sired more than two hundred thousand females and hundreds of males in forty-five countries. (His DNA was even cloned, and two years after his death, scientists created Starbuck II, who, apparently, looked and acted just like the first.)

However, using so much of Starbuck's legendary DNA to reproduce more and more cows could be risky. Too much of one bull's sperm in a gene pool can cause problems. Just as with seeds, the genetic variation in a population narrows dangerously as fewer and fewer genes are passed from one generation to the next. In fact, some fear that this is a problem facing the Holstein genetic pool today. To avoid such a scenario with the Canadienne breed, the association in the Charlevoix is relying on genetic safeguards taken in the past. It purchased Canadienne semen from a private collection and is storing it at the Centre d'insémination artificielle du Québec, which also keeps frozen Canadienne embryos that date back to the late 1980s and '90s; some of the preserved sperm is from the 1950s. These specimens will allow the association to go back in time and capture some of the genetic material that has since been lost; with that they can attempt to broaden the gene pool. But it's not a perfect answer. So far only one in five of the embryos results in a live birth. This means that all the Canadienne embryos that are stored should yield between twenty-five and thirty live calves. If just under half of all births are male—the statistical average—then these embryos should provide about ten to fifteen females that can join the breeding population. Were the association to have more money to invest in the breeding program, it would be able to use advanced techniques to implant dozens of surrogate cows belonging to another breed with pure-line Canadienne embryos so as to speed up the reproduction of the generations. As it stands, it must let nature rule, and it will likely take years to build up the numbers of pure-blood cattle in their herds.

The association's efforts are small in scale compared with what's happening in the Aubrac, where breeding has been a scientific enterprise since the farmers realized they had a problem with their gene pool back in 1978. The Coopérative Jeune Montagne has received financial support from the French government and has constructed a special breeding centre to help preserve the Aubrac cattle. It's a modern building designed to blend into the mountainous terrain, located not too far away from the village of Aubrac where, every fall, the cooperative's farmers evaluate that season's young bulls to see which is the best for breeding. About 160 of the prime specimens are taken to the barn at the back of the building where, over several months, they are fed a prescribed diet and weighed twice a month, their growth noted on a personalized growth chart. In April, technicians arrive to examine the calves, and for each one they make a genealogical tree, noting all the bull's ancestors and listing its qualities. The best two or three males of the year are sent to the insemination centre, where they become sperm donors. The others are sold in an auction. Yet even these bulls are worth a fortune. Buyers come from other countries with similar mountain geographies, such as Ireland, Switzerland, and Germany, and pay an average of $8500 an animal—though the price has been known to rise to as much as $14,000. This selection process has been so successful that, whereas there were only 18,000 Aubrac cattle in 1978, today there are 164,000.

Compare these numbers to the Canadienne and it is obvious the association in the Charlevoix has a long way to go. But Mario doesn't despair. In fact, he is hopeful. He feels that they are in a good position because, for one, Holsteins aren't doing much better as a result of inbreeding. "They don't have more genetic diversity than the Canadienne. They are in a dead end," he said. In fact, he feels the association's work with the genetics of the Canadienne will benefit everyone because it is helping to keep alive diversity in domestic livestock production. Some breeders have recognized that focusing on one gene line has been a mistake and are turning to the DNA of heritage animals for the diversity they need. "The Canadienne has a role to play because she can maintain the diversity of the domestic bovine," he said.

The association in Charlevoix isn't the only group working to preserve the Canadienne breed. Groups such as the Société des éleveurs de bovins canadiens and the provincial government are involved too. In 2000, the Quebec government awarded the breed national heritage status along with the Canadian horse and the Chantecler chicken, other breeds similarly developed in North America. Now the hope is that the provincial government will award the Canadienne a version of an AOC it calls an "attestation of specificity." By petitioning for this kind of legal recognition, the Canadienne association would be following the example of a group of Charlevoix lamb farmers who, under the leadership of a woman named Lucie Cadieux, fought for, and won, an attestation of specificity for lamb.

Cadieux's farm is on the outskirts of a picturesque village in the Charlevoix. Les Éboulements is perched on a mountain overlooking the St. Lawrence River. For decades on her farm, Cadieux has raised lamb for slaughter, meat that long ago became known for its high quality. The animals she raised had adapted well to the climate and geography of the region, and some say that her lamb, as well as that from three other nearby flocks, has a flavour that reflects the terroir. "There's a taste of thyme and hazelnut," said Cadieux. "People say it's a tender meat. Delicate." In the 1990s, the meat became so sought after by restaurateurs in Montreal and Quebec that those who couldn't procure any because of the limited supply would claim they were serving Charlevoix lamb when in fact they were plating New Zealand cuts. Over a period of fifteen years, Cadieux lobbied to have her lamb protected by the certification system so other people couldn't profit from the reputation the farmers had created themselves.

Mario looks to their success and believes it can be replicated. "These animals can provide a livelihood to our farmers," he says. The Canadienne association believes the official designation of their cow will not only preserve the breed but also help to preserve agriculture in the region. The special status will bring value to their work as farmers and, most important, it will bring value to the distinct terroir of the Charlevoix and support small farming communities. Just like in the Aubrac, the farmers are connecting the dots between their culture, their terroir, and the distinct relationship they have with the geography of their area. They demonstrate that this desire to cultivate a new food culture—something we require to carry us into the future—can happen anywhere. And it is.

Terroir and French culture go perfectly together—terroir is, after all, a French word. Yet the idea that food is intimately connected to a place as well as to a culture is more universal than uniquely French. In other countries too, such as Italy, Greece, Japan, and Morocco, people place a similar value on food and understand its relationship to geography and to terroir. Although industrial food often has ruptured this connection, the benefits of promoting a terroir-based food culture are being recognized in many places.

In Morocco and Tunisia, government bodies are creating an inventory of products of the terroir and have already given AOC certification to olive oils, wine, and saffron, among other foodstuffs, to help preserve them and support the agricultural communities that produce them. And in Lebanon, every week in the capital city of Beirut, small-scale farmers gather at the Souk el Tayeb, which is the city's first farmers' market. There they sell their olive oils, goat cheeses, breads, wines, honeys, and regional delicacies such as date- or nut-stuffed *ma'amoul* pastries and *labneh,* a strained yogurt. The market is a grassroots creation, founded in 2004 by Kamal Mouzawak, a food writer and former TV personality who wanted to help people in his city to become reacquainted with their country's agriculture and with Lebanon's rich food traditions. Mouzawak, like so many others who undertake this kind of work around the world, is reacting to the break in food knowledge that people in his country have experienced, the rupture in an understanding that through food we are connected to the land. At base, his effort to build terroir is about reconnecting Lebanon's food culture with the natural cycles of food.

Mouzawak was raised in a village on the slopes of Mount Lebanon. His grandfather and uncle tended to citrus and almond trees, and his father helped them grow herbs and greens and other fruit trees. While he was growing up, Mouzawak's mother spent her days turning the produce they grew into traditional meals. During the better part of his youth, Lebanon was beset by civil war. Looking back, Mouzawak told me that his memories of that period were tied to food, specifically bread. "In war, food is the first thing on your mind," he said. "People's reaction when they heard gunshots was to run to the bakery and stock up on bread." In wartime, the fact that food is central to life becomes clear, a focus we often lose when times are good. After the conflict ended, Mouzawak's path led him to a job as a travel writer, hired to produce a guidebook for his own country. It was while researching this book that he discovered the variety and richness of the Lebanese farming and culinary culture. A few years later he founded the Souk el Tayeb to celebrate this food culture and to preserve its integrity. It would be a farmers' market where producers would sell their own foods directly to the public. The philosophy behind the plan was to support small-scale farmers, particularly those who practised sustainable agriculture—the ones we need to build resilient food systems. "Just the word *food* itself is sustainable," Mouzawak explained of his desire to support people who weren't practising conventional methods. "Sustainable means take and, at the same time, give. If we look at food produced in the traditional way, it's take and give. It makes a whole circle." When we produce sustainable food, we connect ourselves with this whole circle, the cycles of nature. Mouzawak found a way to support this food culture.

Lebanon is a small country, located on the Mediterranean in the Middle Eastern area known as the Levant. This region, as well as its Arabic- and Berber-speaking neighbours, is known for its souks, outdoor markets where trade has taken place for millennia. The souk I once visited in the Moroccan city of Fez, for example, seemed almost suspended in time. Vendors selling goat meat lined the principal roadway, too narrow for a car to pass, so men leading tiny horses that looked to me like ponies and laden with parcels made their way over the cobblestones. When it rained, this road was covered by an awning made from wooden branches, and the smell of grilling meat clung to the dampness. The souk was organized into sections: the dried fruit vendors in one area, the spice sellers in another, the butchers displaying all sorts of meat, including the heads of slaughtered goats, for sale, and then the shoe sellers farther down the path. This kind of marketplace is repeated in cities across the Middle East and beyond. But Mouzawak wasn't satisfied that the traditional model would serve his purposes. He didn't believe that this old-style souk would support the kind of sustainable agriculture he wanted to see thrive in his country. "In a traditional souk, it's the trader who is selling you something he didn't produce. It's like a mall, but open-air. It's like a supermarket," he told me.

So he reinvented the concept, creating an entirely different kind of place where producers sell directly to the customers they meet in person every week at the market. "We cut out the middleman so the money gets back to the producer himself. The most important thing is the close contact between the consumer and the producer. The food is produced by a human being who understands tradition, history, and knows how to farm the land." And this regular contact between the eater and the farmer, Mouzawak believes, will change people's perception of the food they eat. "The consumer has to understand that food is not a commodity that money can buy. Food is something that has to be produced by someone in a field. So the Souk el Tayeb is not a place to buy food. It's a place to celebrate producers." It's a place to reconnect with the food cycles.

Once Mouzawak began, he didn't stop, and the scope of the Souk el Tayeb grew. He founded an organization to help support organic agriculture in the country and started to offer educational programs to promote sustainability. In 2006, he opened a shop to sell the products from the market during the week, when the market was closed. He launched a travelling regional food festival that organizes events in villages around the country, depending on the season. One June day, for example, at the height of cherry season, Mouzawak organized a festival to celebrate the fruit, replete with a farmers' souk, a lunch prepared with local delicacies, cherry picking, and activities for children. In 2009, he went a step further and opened a restaurant in Beirut called Tawlet where, every day, a different producer prepares the foods they typically make at home so people in the city have a chance to taste the regional recipes.

All these projects work to promote the terroir of Lebanon. They have been so successful that people in other countries in the Middle East have heard about them and want Mouzawak to lend them his expertise. He had been helping to develop a Tawlet in Cairo before the revolution in 2011. Qatar's National Food Security Programme has approached Mouzawak to launch a farmers' market, local produce shop, and Tawlet where producers would prepare typical meals and sell local food. However, in these desert countries, the local produce is more likely to come from hydroponic greenhouses than small-scale farms. Whether the model can translate to the more arid countries remains to be seen, but in Lebanon, the benefits are tangible.

"Yes, we lost a lot but we didn't lose it all," Mouzawak said. "People haven't forgotten. People are not satisfied. They are still hungry and thirsty for authenticity and real life." Through sustainable food systems that have echoes of a similar connection to the land—those that exist in the Aubrac, that parallel what Chandrakalabai and other farmers are accomplishing at the organic bazaar, and that value agricultural tradition while adapting to the present in a way that speaks to the experience of the rice farmers in China—people achieve this connection to the natural world they yearn for.

But how does this concept of terroir apply to less prosperous countries where the majority of the people struggle just to eat? It's all well and good for a prosperous society to place a high value on the culture of food, idealizing the close connection we have with nature every time we eat and paying for artisanal products that represent hours and hours of dedicated, painstaking labour, but what about a country where hardship and hunger are the norm? How does the concept of terroir apply in a place where millions of people don't have enough food? In fact, some development experts believe that terroir is just the thing a struggling, agricultural-based country needs to improve food security and fight poverty.

Take Cambodia, for example, where a certification program is helping poor farmers make a decent living and remain on their land. Farmers in the south likely have been growing the Kampot pepper at least since the thirteenth century. It is a domesticated jungle vine that climbs as high as three metres up a pole or a trellis. The pepper plants produce the miniature berries that we grind onto our food for their sharp flavour. Eat the peppers fresh and green, and the small balls have a soft bite. Dried in the sun they become hard, black, and fiery, perfect for adding a kick to your meal. Or peel off the skin before drying to make the rare white pepper, which has a more mellow burn. In the 1890s, the Kampot pepper became an internationally sought-after commodity when the French colonized what they called Indochine and intensified pepper production. Under the French, up to eight thousand tonnes of Kampot pepper was produced each year and exported to France and beyond, where it became known as the best pepper; it is exceptional in steak au poivre. The industry began to slow when the Vietnam War spilled over the border into Cambodia in the 1960s. Then all production ceased in 1975, when the Khmer Rouge took over the country. Over the next five years, two million Cambodians starved to death or disappeared at the hands of the Khmer Rouge and the country's infrastructure was destroyed. The fields that once grew the peppers either were turned over to rice production to feed the starving population or were left fallow, the jungle encroaching on them. After the defeat of the Khmer Rouge in 1979, the country fell into a decades-long civil war. Where there had been about a million pepper poles in the 1960s, during the war, only a few were left, according to the Kampot Pepper Producers' Association. In 1998, the country held a democratic election and some form of stability started to spread around the country. Slowly, the relatives of Kampot pepper producers returned to their fields and began planting again, using the old growing techniques. By 2006, there was enough production that one could declare the Kampot pepper back from oblivion.

With the help of a French NGO as well as a businessman, an association formed and encouraged the government to protect the Kampot pepper with a geographic indication similar to the certification used with lamb in the Charlevoix and cheese in the Aubrac. The farmers wanted this form of protection because other producers had been trying to cash in on their mystique by fraudulently calling their crops Kampot. The farmers argued that their pepper was the product of the terroir where they grew it, affected by the humidity of the region, the regular rainfall, and the porous soils. In the Kampot *cahier des charges,* only two varieties of plants are permitted to be called Kampot and pesticides are not allowed. The rules also outlaw the use of artificial fertilizers, so that farmers would use natural ones such as cow manure and bat guano. In 2008, the Kampot farmers were awarded an AOC. Certification has allowed these farmers to capture the benefit of what their terroir produces. The hope is that they will be able to improve the quality of life in rural areas by making a healthy livelihood earned from their land and their traditional ways. Of course, for this system to work, farmers need affluent people to buy their pepper.

Sustainable food systems don't mean abandoning all international trade. Rather, they are about using trade to support the kinds of farming that are good for the environment and for farmers too. In many cases, long-distance trade in food hasn't helped farmers. In these instances, it would be better to strengthen local food economies rather than rely on international markets. In other cases, international markets can help to support sustainable farming communities by supplying consumers who appreciate the product and who are willing to pay a premium for it. This kind of beneficial trade between regions is important for poor countries—as long as it supports the kinds of farming systems that move us closer to our goals.

A similar hope that terroir can be a solution to poverty exists in Ethiopia. When we think of food in Ethiopia, we are more likely to conjure up imagines of famine than terroir. "Countrywide, we have food shortages and food insecurity. We don't have sufficient supply," said Mekete Tekle, a lawyer and professor who lives in Addis Ababa, confirming the stereotypes. "We want to do away with this problem and we want to be food secure." Tekle is part of an interdisciplinary food security program launched at Addis Ababa University in 2010 that brings together professors from different faculties in the hope that they will help to inspire and educate a new generation with the tools to solve the country's problems. He teaches a course on the right to food and food policy.

Ethiopia is facing a morass of issues. It is a country where population, politics, and policy have come together to create many problems. First of all, the population is growing quickly. In 2011, there were almost eighty-five million people in the country, up from a little more than twenty million in 1960. Most of these people don't have enough to eat. Many farms are too small to yield enough food to feed the family that works the land, explained Tekle, and the modern-style, mechanized enterprises that exist are run by foreign companies and produce food for export. Now the state is trying to join the World Trade Organization, which has brought pressure to liberalize the economy. It supports such private endeavours as the country's largest greenhouse production site, owned by a Saudi oil and mining billionaire, whereby fresh vegetables are grown in Ethiopia and then flown to Middle Eastern cities where there isn't good agricultural land. But this puts the country into the ethically questionable position of being a food exporter with a population that is lacking in food.

The Ethiopian government's environmental protection authority is working to pass a law that will establish the Land Products Centre to take stock of agricultural products and then protect the country's terroir, following the French example. Ethiopia has a rich farming history. It was in this area of the world that coffee was first domesticated, as well as the grain teff. The biodiversity of its food plants is awe-inspiring. Ethiopia already has geographic indication protection for some of its coffee, an international effort initiated by its government in 2005. And the government has taken steps to protect the country's extensive agrobiodiversity too. How terroir-based products can be protected by certification hasn't been worked out yet, and when I was in the Aubrac, an Ethiopian delegation, sponsored by the French government and a French NGO, was also visiting the area so they could witness the effect of terroir and the protection of local foodways. Tekle and his colleagues surmised that in their country they might connect the urban middle class with small farmers who are producing artisanal products in the countryside. A certification system could help market these products and allow farmers to command a higher price. "When you are giving priority to local products, you are protecting the diversity of the local communities," he said. "And so long as they continue to get the benefits, they will preserve that. The purpose of the program is to encourage them to continue."

Terroir as a development strategy isn't straightforward. There are inherent problems with the approach. For example, in Kampot, farmers are so poor that they often aren't able to capture the benefits of the certification because their vines require years to mature, and the farmers are often in need of an immediate source of cash to feed their families and aren't able to wait for the high returns. There is also the question of how to connect small artisanal farmers with the middle class and elites in the city or in other countries who will pay the higher price for the product. When small farmers in the developing world start to sell their product on the international market, they become part of the worldwide economy of food, dependent on the ebbs and flows of international prices—after the global financial crisis of 2008 and 2009, the market for Kampot pepper shrank. When there is demand for the goods, trademarks must be protected by the law to be effective. Farmers need the help of a proactive and litigious government that is willing to prosecute those who try to profit from protected domestic trademarks, a role the French government plays well. This type of certification is covered by two WTO agreements, but there must be a will at the government level to enforce them too. And then there is corruption. While promoting terroir could help to increase food sovereignty in a country like Ethiopia, many doubt that the poor farmers in Ethiopia's current democratic climate will be the ones to benefit. In Cambodia, in the Kampot area, it was reported that a senator took nearly four thousand hectares of land that was well suited to growing the pepper and turned it into a plantation for sugar cane, a commodity that can earn quick cash.

So it remains to be seen whether this certification and marketing of terroir can truly help people in developing countries while protecting their agricultural way of life. Their well-being, however, is of critical importance to the success of sustainable food systems.

Protecting the cultural diversity of our food has another worth to society beyond helping the farmers. According to the documentation from the French institute that oversees AOC certification, protecting the integrity of traditional products, made by hand, is about preserving a collective heritage. It is about ensuring a lasting farming system, not only because of the tangible benefits to society today but because there is an intrinsic value in food culture. If food culture makes us human, then a diversity of food culture distinguishes us from other humans—it defines who we are. How I mix wheat with water and yeast and then bake a loaf in a wood-fired oven differentiates me as a cultural being from the person who mixes flour with these same simple ingredients but then rolls the dough flat and fires it on the glowing red side of a hot tandoor.

The significance of these seemingly small differences was underlined for me when I visited an Afghan bakery in the Toronto suburb of Scarborough. A team of bakers worked the dough in a windowless back room of a strip-mall grocer where they sold everything from Persian carpets to canned fava beans. They were trying to replicate the traditional naan that typically would be made in the communal oven of an Afghan village. One of the managers gave me a tour. He explained to me how the bakers used the tips of their fingers to make rivulets in the bread to achieve the perfect texture, and he showed me the gas-heated oven about the size of a cube truck that they'd fashioned with rotating metal shelves in order to mimic the tandoor's cooking process. They were making a big effort to replicate this way of making naan in Scarborough because of its significance to Afghans living there. "This bread is what makes us Afghan," the manager said. "Without this bread, I am not Afghan."

But the global industrial food system works to minimize these differences. It wants to sell the same bread to people in Kabul, Dallas, Hong Kong, and Cape Town—bread that is divorced from local cultures. For terroir scholar Amy B. Trubek, the prospect of a globalized food culture that is severed from terroir and where everyone eats similar foods is horrifying. "Should we all be the same? Should we eat exactly the same thing, delivered to us in exactly the same way?" she said. Preserving local food traditions is therefore a rejection of the global industrial food system.

The cultural diversity of food goes even deeper. It's part of our genetic heritage. We've evolved as a species to have it. There is a brand-new field called evolutionary gastronomy that blends anthropology and science. One of its pioneers is an academic at the University of Arizona named Gary Paul Nabhan, an ethnobiologist and plant ecologist who lives on a mountain orchard where he grows some forty-five varieties of desert fruits. (He is also the author of a book about the Russian biologist Vavilov. Interestingly,

Vavilov too made the connection between biodiversity and cultural diversity.) In Nabhan's groundbreaking book *Why Some Like It Hot*, he tells the story of how the foods we eat are a reflection of an interaction over the millennia between biology and cultural diversity. He argues that we are not only what we eat, we are what our *ancestors* ate.

The culinary cultural diversity that we see around the world today started long before the dawn of agriculture, when different hunter-gatherer communities would have made decisions about what to eat, decisions that put these various ethnic groups on their own particular evolutionary courses. During prehistory, people needed to distinguish what was good to eat from what would make them sick, or worse, and therefore they developed social mores and norms about food to guide them. "Their favourite tastes, textures, and colours has had a huge effect on why today we have an enormous diversity of tomatoes, for example," Nabhan explained to me. Humans guided evolution through cooking and eating. "We changed the genetic makeup of crops based on how we liked their flavour for cooking." These foods would have been chosen for how good they tasted, or how they made people feel. What people ate also exerted an evolutionary pressure back on communities. "We selected foods for colour or texture, but that affected their nutritional quality." Nabhan concludes that while decisions about what to eat may be cultural, whether or not that choice satisfies basic nutrition is biological. And through evolution, there has been an interplay between our genes and the food choices our ancestors made. Such as with spices.

In 1999, Paul Sherman, an evolutionary biologist at Cornell University, published, with Jennifer Billing, a research paper that investigated the age-old notion that people who live in tropical countries, where temperatures are the hottest, have the spiciest cuisine. They reviewed dozens and dozens of traditional cookbooks and analyzed recipes from thirty-six countries. They found that in fact it is true that the closer you live to the equator, the spicier your food and that the spice use isn't random. By preparing foods with these flavourful additions, people were protecting themselves from food-borne microbes that flourish in hotter climates. When added to a recipe, herbs or spices such as oregano and allspice as well as garlic and onion inhibit bacteria growth or flat out kill them before the germs have a chance to make us sick. "The microbes are competing with us for the same food," Sherman told *Science Daily* at the time of publication. "Everything we do with food—drying, cooking, smoking, salting or adding spices—is an attempt to keep from being poisoned by our microscopic competitors. They're constantly mutating and evolving to stay ahead of us. One way we reduce food-borne illnesses is to add another spice to the recipe. Of course that makes the food taste different, and the people who learn to like the new taste are healthier for it." The authors also pointed out that those families who cooked with spice would probably have had healthier children who would in turn have grown up to eat spices, passing this beneficial food culture from one generation to the next.

If our DNA reflects a diet that grew out of the climate and geography where our ancestors lived and ate, then the homogenized, global diet of the industrial food system is bad for us at a physical level too.

"The Western diet is becoming the dominant diet at tremendous cost, not just because everyone is vulnerable to junk food, but because there are protective foods in all of those diets that match with the genetics of those who ate them," said Nabhan referring to the diet that is high in saturated fats, sugars, and refined grains that is so common in the West. "There is no optimal diet for the incredible cultural diversity of people on this planet." Not only does a diversity in food culture taste good, but we need it to survive and be healthy. And for that diversity to thrive within the context of today's industrial food system, we need to value the terroir and then reconnect to it. With international migration and mixed marriages blending cultures and gene pools, we won't likely return to a world where most people eat a diet similar to that of their ancestors. However, certainly if we continue down the path of worldwide consumption of pasta, hamburgers, and fried chicken, it would be bad for our health—and for our souls.

"If we were all the same, I think there would be absolute sadness," said Eric Barraud. Barraud is a tall forty-something Frenchman who grew up in a village not too far away from Monsieur Valadier's Aubrac. He founded an organization that he calls Terroirs et Cultures to preserve this diversity of food culture. Barraud has witnessed how, when terroir is valued, diversity thrives. His organization works in particular with countries around the Mediterranean, but also advises others in the field. It hopes to reverse the trends of globalization whereby many food and agricultural traditions are disappearing. "Terroir brings biodiversity, cultural diversity and food and human diversity," Barraud said. The Aubrac is one of its star examples, and Monsieur Valadier sits on the board of directors. Barraud regularly brings delegations from other countries such as Lebanon and Turkey to learn from what has been achieved in the region. The hope is not that they replicate exactly what's happened in the Aubrac—again, it's not about homogenization—but rather understand how the farmers in the Coopérative Jeune Montagne identified what they had that no one else had (their terroir) and use this to foster the culture of terroir that supports a life and a landscape. "Terroir can't be photocopied," Barraud said. "But you can be inspired to make your own."

It took two hours to follow the cows up the mountain to their pastures as part of the transhumance that day in the Aubrac. Once we arrived at the summer pasture, the farmers got to work corralling the animals in a pen to remove the cowbells they didn't need anymore. The rest of us made a circle around the herd to keep them in one area. When the task was complete, we were just as happy as the cows to have reached our destination and wanted to eat too. The group of humans made its way to the old *buron* that the Valadier family had fixed up and now used as a cottage, with a handful of beds upstairs, a big room with a fireplace below, and a makeshift kitchen at the back. It's a tradition for Monsieur Valadier and his sons to prepare aligot after the transhumance to feed the guests. In the back room of the *buron,* they got to work, boiling an enormous pot of potatoes. Meanwhile, the glasses were brought out and the rest of us gathered outside, around a spring bubbling from the earth, and quenched our thirst with diluted *gentiane,* the brilliant puce-coloured liqueur made from the native flower that grew in the grasses around us. We then ate a traditional Aubrac savoury cake made with prunes. But everybody was waiting for the aligot.

I went inside with some of the others to watch. Monsieur and his sons were stirring the mixture with a large spoon the size of an oar, and worked their arms to blend the thick mass. Finally, when they'd stirred for what felt like half an hour, it was time for Monsieur Valadier to do the honours of checking to make sure the aligot was ready. He dipped the oar into the pot and then lifted the aligot over his head. The string didn't snap. The aligot was perfect.

On my last day in the Aubrac, I went to the Valadier farm to watch the evening milking. Both in the morning and at the end of the day, Monsieur Valadier's sons Géraud and Jean gather their herd of milking cows in the nearby fields and walk them, with the help of their dog, to the dairy barn in the hamlet of Les Clauzels. When I arrived, the sun was beginning its descent and the late-spring fields glowed under its orange rays. The sound of the insects and birds calling was so loud, I could hear them through the open window of my car over the rattle of pebbles flying against the chassis. It was a perfect evening.

It was quiet in the barn. The sunlight slanted in through the side window. There was the sweet smell of manure and of water on the concrete floor. Géraud had brought the cows inside and was hosing down the mechanical milking equipment before he began, which he said would make it easier to clean afterwards. Géraud is thirty-eight, the youngest of the Valadier children. Unlike his dad, he is slight, with curly hair and a mild, introverted disposition that seemed to put the cows at ease as they waited silently in a queue for the milking to begin. Then he opened the gate and the first dozen ladies—they really looked like ladies with their long eyelashes and soft jawlines—walked in a single file that effortlessly split into two as they took their positions on each side of the equipment. The way they moved into their spots reminded me of a well-choreographed ballet. Géraud used a handful of wood shavings to wash their teats—a method he'd learned from the Swiss that is meant to preserve the cow's skin and use less water—and squeezed them by hand to make sure they weren't infected before hooking them up to the pump. The other cows watched serenely and waited for their turn, resting their heads on the gate or on each other's backs. "I think they like it," he said. "When you know the cows you are milking, it's a pleasant job. I know before the milking how much milk each will give. If they don't, I know there is a problem. There is a certain pleasure in seeing a milking run smoothly. It's like a chef having a meal appreciated."

Géraud and his brother used to have other jobs. Géraud was trained as an economist. He worked for the French foreign affairs department and was posted to the consulate in Montreal for a year. But after he was married, he returned to the farm to join his older brother, who'd already come home from his job in the city. They both wanted to farm, and the plan was to take over the daily responsibilities of the cows from their dad when he retired. "When I was in school, I realized that I was drawn to the earth. I did lots in the city, but I realized that I didn't feel comfortable there." And because of his father's life work to build a food system that is based on the terroir of the Aubrac, on the culture, Géraud had something to come home to. "Everything that was done, with the Aubrac breed of cattle, with the Laguiole cheese—all this has made a viable life and has allowed young people to settle here." If young people are the future, and culture is a vehicle to supporting a sustainable food system, then the Aubrac offers an example we can all learn from.

Despite their success, they are not resting on their laurels. Monsieur Valadier has plans for a whole lot more—a whole lot more than even he can accomplish in his lifetime. When I spoke with him after I returned from my trip and asked him, when he answered the phone, how he was, he said he was well, but he'd be a lot better if he was thirty years younger. Monsieur Valadier still has a lot he wants to do. But now the next generation is involved. Géraud and Jean, with their dad, are trying to bring back the Aubrac breed's dairy capabilities. The family owns five of the one hundred or so cows that farmers in the co-operative are breeding to favour milk production. Their dream is that one day they will be able to make Laguiole with the milk of the Aubrac breed once again. And then there's the idea that would crown Monsieur Valadier's success. He has started the process of having the government create a national park in the Aubrac, an area where the terroir would be protected in perpetuity. "I want this land to remain a resource for everyone to share," he said. "I can't think of a better way to move into the future."

Efforts to preserve terroir are for the common good. Just as we all benefit from biodiversity in agriculture, so do we all gain from diversity in terroir. The question for the future is how to best transmit this culture and support the kind of food system we need in 2050.

CHAPTER FOURTEEN

Introducing . . . Food: The Culture Shift

In an elementary school cafeteria in Paris's 2nd arrondissement, a group of ten-year-olds sat down for lunch. In some ways it was just like school lunch hour anywhere: the kids became rowdy, the teachers raised their voices and told them to be quiet, the kids calmed down for a few minutes before their voices started to rise again. But mostly, this was a school lunch unlike any I'd seen before. It wasn't because the school was located in central Paris, a few blocks from the old stock exchange, across the street from the covered shopping arcades of the Belle Époque. Or because the cafeteria, located in the school's basement, looked distinguished with the stone arches of its eighteenth-century foundation dividing the room. What was different about this lunch began with the food: endive salad, boeuf bourgignon, and *flageolets au thym.* The children sat together at round tables, eating good-tasting organic meals—mostly locally sourced—off real plates using real cutlery, wiping up the juices with baguette made from unbleached flour that came from wheat grown nearby in Isle de France.

This was a stark contrast to my own school lunches in Canada. I attended a public French-language school in Toronto that was a bus ride from my home. So my mom packed me a lunch and I joined the rest of the kids every noon hour in the auditorium to eat. I can still smell the acrid cleaner they used to wash down the long tables that were folded out for us every day. This, mixed with what smelled like Wonder bread and spilled apple juice, greeted us as we filed in. I hated lunch. I wasn't a picky eater, but nevertheless, there was nothing I was happy about eating at school. One of my preferred meals was a grey, waterlogged hot dog, floating in a thermos of warm water, that I would fish out and place on a bun that typically had been squished in my bag. (This was the 1980s, when hot dogs had a better reputation than they do today.) My mom saved the little ketchup packets from fast-food restaurants to send with me. On special fundraising days, volunteer mothers would come to the school kitchen and serve us freshly boiled hot dogs or ordered-in pizza. Lunch meant fuelling up with enough food to get through the rest of the school day.

That's generally the way we approach food when it comes to kids: food is fuel. We think about feeding our children as coaxing them to eat enough to fill their stomachs until the next meal. Our approach to school lunch is indicative of this. In the United States, the national lunch program has been feeding children the excesses produced by agricultural subsidies for decades. The US government established the National School Lunch Program in 1946 (though volunteers and charities already had been feeding kids at school for years), and they offloaded some of the extra food that farmers were churning out after the Second World War. Today that same program feeds more than thirty-one million children in one hundred thousand schools, but the lunches have been notoriously poor, serving up high-fat, high-salt processed food. In 1981, when President Ronald Reagan's administration cut $1 billion from school lunch programs, it was proposed that if ketchup and relish could be reclassified as vegetables, the condiments would count as a food group and fresh veggies could be struck from the menus, lowering costs. (The suggestion didn't go forward after a swell of protests.) More recently, people have decided that bad school food isn't good enough for kids and many people are trying to improve the foods that are served in educational settings. However, there is still a general lack of awareness about the importance of what kids eat at school, let alone any inkling that the culture of terroir should play a role in these meals. And as long as this situation persists, it should be no surprise that many kids grow up to make poor meal choices as adults.

How strange it was to learn that in France—not only in the Paris school I visited but in schools all over the country—feeding kids well and teaching them an appreciation for good food and its terroir is a priority. School lunch programs, administered at the municipal level, provide a meal in the school canteen for six million children a day. A growing number of these canteens have chosen to serve primarily organic food, as was the case at the school I visited in Paris. Laws stipulate what children must eat across the country, and these laws are revised and reformulated with changing attitudes. A generation ago, school food in France didn't have the same reputation it does today. According to one Paris chef in his thirties whom I spoke with, school lunches when he was young were less than appetizing. Then parents became involved, the government acted, and the school canteens improved. In 2011, the state outlawed most fried foods, as well as children's unlimited access to ketchup and mayonnaise, and specified how many times a week schools had to serve meat, fish, and eggs. To make sure the lunch-hour meal is available to everybody, the state funds the program. However, families are asked to contribute an amount determined by their income. Consequently, at school all children can eat together and explore a wide variety of tastes at the same time as they fill their stomachs.

Providing children with a school lunch is, on the one hand, about ensuring that kids are well fed. There is also, however, an element of socialization. French society has made a decision to support the taste education of the next generation by making efforts to preserve the culture of food—including banning ketchup from being served with traditional dishes at the school canteen. Every year since 1989, a non-profit has organized a countrywide week-long program called La Semaine du Goût—National Tasting Week—with the purpose of educating children about taste, terroir, and food culture. Chefs and culinary professionals volunteer their time to go out into the nation's classrooms to offer cooking classes, tasting sessions to develop the palate by learning to distinguish different flavours, and lessons about where food comes from.

The belief in the importance of educating children about these traditions is also seen in the family. "The tradition of preparing food is part of our culture," said Benjamin Locreille. "When I was a kid, I learned to cook with my mom and my granny." Locreille is the director of a cooking school run by Michelin-starred chef Guy Martin where they teach children all about cooking: how to use knives, how to pan-fry fish or make escargots and macaroons. It's only one of the many cooking schools in Paris where parents send their children because cooking food is integral to their identity. (Martin's school also offers adult classes.)

"The art of preparing food is something like literature that is a part of our heritage," said Locreille. "Here my role is protector of our heritage. It's something that can be passed on." The children who come to the school receive their lessons in a professional environment; in the classroom kitchens, there were gas ranges, long stainless steel work spaces, and, hanging on a magnetized strip on the wall, dozens of sharp chef's knives. When I remarked about putting these potentially dangerous instruments in small hands, Locreille explained that the children must be old enough to handle a knife before they can enrol in a class—that is, he said, they must be at least six years old.

I laughed at the thought of American six-year-olds being given chef's knives. "I've never had a problem with the knives," said Locreille, a little incredulous at my response. This to me underlined the starkly different relationship that exists in France between children and food. In North America, we are fearful when it comes to mixing kids with food—fearful of knives, certainly, but also worried about what we perceive to be the tyranny of the child's palate. To cater to these tastes, we have created kids' foods that are sweet and colourful and come in interesting packages. Bright pink yogurt is sold in a plastic tube kids can suck on. Parenting magazines recommend that mom and dad encourage their kids to eat by making food fun, such as using a cookie cutter to create animal-shaped sandwiches. And in restaurants, children's menus offer "kid food" such as pizza or chicken fingers. All this panders to a preconceived notion of what kids will like *and* consume and limits our children's experience with food. Just as we wouldn't teach knife skills to a six-year-old, no one would put escargots on a children's menu where I live. In France, teaching a child to enjoy eating snails is simply part of a good education. "It is a cultural inheritance," said Locreille. "It is passed from generation to generation."

His belief demonstrates that this connection to food—this connection to terroir—isn't genetic. It isn't something that babies are born with in France, or anywhere else. Rather, it is something that is taught, a social project. In 2010, UNESCO declared French gastronomy to be part of the world's intangible cultural heritage. This designation is given by the global body to acknowledge that the heritage of the world's many cultures is not limited to the concrete, such as the Taj Mahal, but rather that every culture has a legacy that is just as important to protect for posterity. After the honour was announced, the French government responded with a statement indicating that the country would take strong action to ensure that this heritage endures. It would put its focus on schools.

Other countries have taken a similar approach to educating children. In Italy, the government invested more than $260 million in school meals between 2004 and 2007. There, schoolchildren are seated at proper tables set with a real tablecloth, plates, and silverware. Everyone is served a three-course meal cooked in the school. The meal is often organic, sourced from about four hundred Italian organic farms. The menu of one public elementary school in Rome reads just like one in a restaurant. Every day the cook posts a handwritten note telling parents what their children will eat at noon. The kids always start with a *primo,* a pasta dish such as spaghetti with tomato sauce or ravioli, and then are served a *secondo,* a meat dish that could be chicken or beef. Next, they eat an *insalata,* covering off the vegetables, and finish with a *dolce,* dessert. Children, the Italians recognize, are the connection to the future of food.

Here, there is a growing interest in food education for kids. In Berkeley, California, Alice Waters of the world-renowned restaurant Chez Panisse helped to found a program called the Edible Schoolyard Project at Martin Luther King Junior Middle School. The program not only teaches kids how to grow food on a half-hectare plot and then prepare it in a learning kitchen, but also uses the garden to help teach traditional subjects such as science and math. Today the project operates at nearly 680 locations. This approach has been repeated across the United States and in Canada too. In my city, Toronto, more and more schoolyard food gardens are offering children the chance to learn about how food grows and to make that connection to the earth. "Working with kids is very exciting because it's hopeful," said Debbie Field, the executive director of FoodShare, an organization in Toronto that works to build a better food system in the city. It has created all sorts of programs that involve people with their food, from running community gardens and compost programs, to buying fresh fruits and vegetables in bulk to sell at a fair price to school meal programs and offering fresh fruit and vegetable boxes to the public at an affordable price, to renting space to small start-up food-based businesses. But a cornerstone of its work is education. "Kids are very receptive," said Field. "Kids are important because they are the conveyor belt of all change."

This is one reason that FoodShare advocates putting food literacy on the school curriculum and formalizing the transfer of knowledge. "It's time for the public education system to heal the crisis of food," said Field. She wants all children in school to be taught to cook, to grow food, to compost, to keep bees and chickens, and generally become food literate. This, she says, will arm children with the know-how they need to feed themselves and keep healthy in the future. "Curriculum is about how we train a whole new generation in habits—whether it is computers, food, reading, or math," she said. "I am strangely optimistic. I think we're at a tipping point."

And in the future—a future where we have a more just and sustainable food system than the one we have today—Field believes that we as a society will look back and see that it was the change in the schools that triggered the transition to a sustainable food system that can feed us today and tomorrow too.

But is such a cultural shift possible? Even in France, with its widespread understanding of terroir and deeper connection to food, the industrial food system is well entrenched. After all, industrial agriculture is alive and well in the country; the French company Carrefour is one of the world's largest supermarket chains, with operations in more than thirty countries. Even those who have been reared on an appreciation for French regional specialties buy processed foods from companies such as Picard Surgelés, a French agri-food business that makes frozen meals of all sorts and runs its own supermarket chain where it sells nothing but house-brand frozen products such as guinea fowl stuffed with figs and chestnuts, Vietnamese soups, and molten chocolate cakes. According to a report written by the American embassy in Paris for investors interested in the French market, a shift in food culture is under way. The French are increasingly likely to eat out of the home, which is changing the market in food.

In many ways, we're all far from a culture of food sustainability, no matter where we live. But in other ways, some of us have already started to reconnect to the cycles of food.

I arrived in Detroit by crossing the bridge over the Detroit River, and my first sight of the city was the shell of the abandoned central train station. The building, constructed in the early 1900s and abandoned in the 1990s, appeared ghostlike, grey and melancholy, standing two dozen storeys high with floor after floor of broken windows, surrounded by patchy grass and cracking pavement. I continued to the downtown, which was only a kilometre farther, to find more of the same. While the centre of town had experienced a bit of a renaissance, there was little traffic on the road; only the odd person walked down the sidewalk, and there were many boarded-up shops, empty lots, and more vacant buildings that looked like they once had been grand but had long been forgotten, like an old luxury car left out to rust.

I'd come to Detroit because the city is legendary for its urban agriculture. Journalists had flocked here to tell the story of how residents had turned the backyards of burnt-out houses into farms and how community groups were tending fields and fields of fresh vegetables. Motor City was a mecca of urban agriculture. I may have come for the city-grown vegetables, but I left with an entirely different story that surprised me. In Detroit I learned that urban agriculture is about more than growing food in the city. It's about that cultural shift, the reconnecting to the food cycles. In Detroit, I witnessed the transformative nature of food and the speed at which people can change their food narrative, their food culture—a cultural transformation of the kind we must all undergo soon.

I'd heard that things were bad in Detroit, but I couldn't fathom how bad until I arrived. According to the 2010 census, more than a third of the population lives below the poverty line. The city lost 150,000 auto industry jobs between 2000 and 2008. The entire state of Michigan lost 44 percent of all its manufacturing jobs in the first decade of this century. Most people haven't found good alternatives. The population of the city continues to drop as more of those who can leave move elsewhere. What this means is that the city is lacking in many ways. For one, there are no national supermarket chain outlets in Detroit. Not one. The last two supermarkets, both owned by A&P, shut down in 2005. There are small grocers where you can pick up milk and basic products such as canned foods, and on Saturdays there's a large farmers' market downtown, but shopping at the supermarket like most Americans is no longer possible. This makes the entire city of Detroit a food desert. "A large part of the population relies on the corner gas store, the corner party store, the liquor store," said Kwamena Mensah, manager of the Detroit Black Community Food Security Network's D-Town Farm. "The food you get in those places is going to be high in salt, high in sugar, and high in fat—processed foods. But people have to shop in those places because there are no options." When I stopped to buy a map at a corner gas station, I stood in line behind a young mother with a child in a stroller, waiting to pay the man standing behind a wall of bulletproof glass. The man spun her change back through a small rotating flap in the glass, something I'd only ever seen on TV. "If you have transportation, you go to the suburbs," said Mensah.

But most people in Detroit don't have transportation. My first stop was the Earthworks Urban Farm, founded in the Island View neighbourhood by a Capuchin monk in 1997 as an extension of the soup kitchen the monks ran. The afternoon I pulled up, the staff was holding a clinic where people in the community could come to repair their bicycles; just as there aren't supermarkets, nor is there a bike shop nearby. Patrick Crouch, the farm manager, met me. "It's a difficult bike culture," he explained. "It's not a choice, it's a necessity for most people." They ride their bikes because they can't afford a car, and there's no reliable public transportation. (Buses don't arrive when they are expected.) And without a car to navigate the web of freeways that connect neighbourhoods, they can't get to the suburbs to visit a supermarket. "The social fabric in Detroit has largely been destroyed. There are jagged teeth of neighbourhoods at this point. You are lucky if you have an incisor. The neighbourhoods have fallen apart, but the churches are the one thing that has kept the community together. Churches and barbershops are the two places Detroiters come together," said Crouch. And increasingly in food-producing gardens too, it appears.

The Earthworks Urban Farm is spread over a few blocks. It has vegetable gardens, a youth garden, greenhouses, a community orchard, and beehives. The food it grows supplies the soup kitchen still run by the monks, and it also sells the food at farmers' markets set up by Earthworks, where people can exchange their government food coupons for certified organic vegetables. The Capuchin monks have been in the area since they built their monastery in the early 1880s, on what was then farmland on the outskirts of the city. Over time, the neighbourhood around them changed with Detroit's fortunes. At the turn of the twentieth century, Detroit was a boom town, a prosperous place because of its position on the Great Lakes transportation route. There was a big lumber industry in the city, ironworks produced stoves, and there were also steel mills and ship building. "This neighbourhood tells the story of Detroit's economics," Crouch told me as we stood in the vegetable garden a block and a half from the office. "This building right here was headquarters for a lumber company. The garden was the lumberyard where they milled and cut lumber. Down the road was the largest stove factory."

By 1914, Detroit was the Motor City, named for the importance of the automotive industry. The big three car companies—the Ford Motor Company, General Motors, and Chrysler—all had their headquarters in that part of Michigan. By the 1920s, the city's auto industry employed more than seventy-five thousand workers, and people from all over the United States came in search of good jobs. During the Second World War, the car factories were temporarily converted to manufacture military equipment, and industry continued to do well. Then the city's fortunes began to change. Racism had long been an issue. Before the 1940s, Ford was one of the only companies that hired African-American workers and then only for hard, manual jobs. According to the book *The Origins of the Urban Crisis: Race and Inequality in Postwar Detroit* by Thomas J. Sugrue, as more and more people from the southern states moved to the city and the African-American population grew, the white workers who had benefited from the racist hiring policies perpetuated the inequalities and racial divisions. The city became "wracked with racial tensions and conflict," in Sugrue's words. In 1967 a police raid of an unlicensed bar led to five days of civil unrest that is known both as the rebellion and the Detroit riot. Hundreds of people were injured, dozens were killed, and more than twenty-five hundred buildings burned. After it was over, more and more of those who lived downtown moved to the suburbs, many of them white families.

To this day, Detroit is a racially divided city. According to the 2010 census, the larger metropolitan area is one of the most segregated in the country. The week I was there, the local news reported on a bus trip that was organized to take predominantly white people who had spent their lives in the suburbs of Detroit to see the downtown they had never visited.

When manufacturing jobs started to quickly disappear in the 1980s, more and more people left the city. The population has dissipated so much since then that there were more than 60 percent fewer people living in the city in 2011 than there were in the 1950s. Neighbourhoods have been left hollow, with street after street of empty homes. Detroit's high foreclosure rate has also contributed to the high number of boarded-up properties. And when properties are vacant, there's no one to pay the property taxes and the municipality can't afford to maintain services. In 2012, the city didn't have the money to fix the 40 percent of its streetlights that were broken. Fire stations have been closed, and thieves have stripped them for their scrap metal. Often, empty houses are broken into and set afire, which is why on a drive around town you see so many burned homes. When David Bing was elected mayor of Detroit, he pledged to demolish ten thousand buildings before the end of his term in 2013. This is why there is so much room for growing food in the city.

There are whispers of a downtown revival. The city is investing in refurbishing the waterfront, and some of the old disintegrating buildings are being turned into condominiums. A twenty-something told me that his apartment building in the city centre is now fully rented and the landlord is raising rents. A number of locals attributed the change to the arrival of young people from other cities looking for cheap real estate and the chance to grow food in a city. But this is causing new tensions too. Mensah told me, "There is a lot of resentment because a lot of times people feel the administration"—

the city—"is catering to these young whites coming into the city." It's not only in Detroit where food movements are recognizing that addressing racial inequality is an important part of building sustainable food systems because race inequalities—as well as class—extend throughout the food system. For one, people of color in the United States are disproportionately affected by food insecurity. According to a report by the Applied Research Centre, a national racial justice organization, both black and Latino families experience food insecurity at three times the rate that white families do. Food related disease also skew for race. The same report holds that blacks have the highest obesity rates in the country, followed by Native Americans and Latinos. Also, the low income jobs in food from the farmers' fields all the way down the chain to the low wage service jobs of the sector are frequently held by people of color and immigrants. ARC for one is calling for an increased awareness in the good food movement. Urban agriculture in Detroit takes place against this backdrop.

A lot is happening in the city. In addition to Earthworks, there are other community-minded farms, such as the D-Town Farm, where the Detroit Black Community Food Security Network grows vegetables on a plot of parkland that it leases from the city. There are also many community gardens and vegetable patches in different neighbourhoods such as Brightmoor, about twenty-five kilometres from downtown. As I pulled into Brightmoor, a sign by the side of the road welcomed me to the "Brightmoor Farmway." This name didn't exaggerate the scale of what's happening.

The Brightmoor neighbourhood was home to an astounding number of food-producing gardens. A woman who lives there, Nikki Eason, offered to take me on a tour, and even she was surprised by what we came across on our walk. Up the road from her house were a potato patch and at least two seed-saving gardens. At one intersection, three out of four corners were producing food. A sign at another patch of land read "Ms. Gwen's Edible Playscape." "I see someone is trying to start a garden there," said Eason, pointing to yet another parcel. "They'll start them anywhere." We came upon a group of young adults working in a garden, weeding between the rows. One of the men told her, "We're just learning." "We are too," she answered. She told me, "A lot of kids running around here don't have nothing to eat. That lunch at school is their only meal. The food around here is going to help a lot of folks."

Eason had lived in her house for seventeen years, after inheriting the property from her dad in 1994. It was a nice neighbourhood back then, she remembered. But a few years after she moved in, the area started to crumble. "People started to throw trash around," she said. Properties were neglected, and the weeds and debris covered the sidewalk near her place. "You couldn't even look down the street and see the sidewalk. At first I wanted to move because I was sick of the nastiness," she said. Then a woman named Riet Schumack moved into the neighbourhood and started to organize the community, founding an organization she called Neighbors Building Brightmoor. She planted a vegetable garden and started the Brightmoor youth garden. The idea was contagious, and more and more people followed her example, including Eason. "When I saw what they are doing, I told my husband, 'I want to sign up.'" They've also raised money to fix up the local playground. They've cleaned the sidewalks and painted the boarded windows of abandoned houses with colourful, happy images to help disguise the blight.

When I was there, Eason had recently begun to tend beautifully straight rows in an abandoned lot a few doors down from her place. Her plan was to grow enough food to sell the extra. The following summer she told me she took her cilantro, tomatoes, and other produce to market. She followed the example of other people who have been doing this already through an organization called Grown in Detroit, which aggregates city-grown produce and sells it at the farmers' market on Saturday mornings. "Ghetto, that's what they call it. But I just call it middle class," she said of her neighbourhood. "My mom said, 'Don't give up on Detroit. It's going to come back. Just be patient.'"

Such is the transformative nature of food. And that's the point in Detroit. The way people are using barren lots for food production can't be replicated in the growing megacities of the world—there just isn't any space. And no one believes we should be finding room in our packed cities to grow all our food, particularly the grains and the crops that require vast tracts of land. Rather, it's the cultural shifting that is related to urban agriculture that we can learn from. In Detroit, growing food provides fresh vegetables where there aren't any supermarkets and helps the urban farmers, who have extra produce, to earn a little money. Growing food has also empowered people to make changes to their neighbourhoods. And there is a mindfulness of the significance of what is happening in the city that goes beyond how many pounds of vegetables people can produce. "The corporate model has failed us and the answers are coming from the community," Kwamena Mensah of D-Town Farm told me. At Earthworks, a similar perspective exists. "There is something deeply spiritual about growing food, eating food with other people," said Patrick Crouch. "We are an agrarian people. It's only a generation or two that we are removed from that. It's a connection to our roots."

The name of the Earthworks farm speaks to the broader significance of urban agriculture and the goal the group has to reconnect people to the food they eat, to mend the damage wrought by industrial food. An earthwork is a military term describing an earthen barricade designed to protect from attackers. The farm has applied the word to the idea of environmental protection: its sustainable farm is an earthwork that protects the natural world. The farm also imbues its name with a more literal second meaning. They say the earth is working when it produces food and reason that we humans are the shareholders in a living system. Through urban agriculture, people can begin to reconnect with this cycle because when they cultivate food, they can feel the soil, smell the earth after a rain, watch the earthworms, witness food grow from the land and understand where their nourishment comes from. In this way, a cultural shift can begin.

This reconnection is happening in cities everywhere. The proof is in the rise in the popularity of backyard chickens and the growing number of urban beekeepers as well as community gardens, front and backyard vegetable gardens, and allotment gardens. All over the world people who live in cities are connecting with their food—with nature—through urban agriculture. In New York City today, there are more than a thousand community gardens on city land, and in Montreal, more than ten thousand citizens head to eight thousand allotments and ninety-seven community gardens to tend to their food plants. In 2013, a program was announced by the Chicago mayor's office to hand over vacant city lots to Urban food production. In Berlin, eighty thousand people are involved in urban agriculture. There is a rising interest in growing food on balconies and terraces in Bangalore, where a local radio station has held call-in programs with an organic farming expert who answers people's questions. In Beijing, a food activist who throws seed bombs took me on a tour of a hutong, one of the city's old neighbourhoods of narrow streets and one-storey compounds, to show me all the places people were growing food. Although there is no official count of urban farms in Hong Kong, people say there are hundreds, including one rooftop where an entrepreneur rents out small plots to gardeners. In the dense urban neighbourhoods of Paris, people come together to grow vegetables in the *jardins partagés*, gardens that are shared by members of the community that range in size from the tiny to the small. As one gardener told the writer of an academic paper on the topic, the *jardins partagés* "are highly symbolic places where people can re-connect with Nature, in a magical and surreal context, between buildings and asphalt."[48]

In the United States and in Canada we see a growing movement to involve municipalities in urban agriculture in a novel way, by using public land to grow food for whoever wants it. Rather than planting flowers and ornamental shrubs and trees in public space such as city hall gardens, school grounds, and hospital lawns, the idea is for the government to make room for food plants and trees. The concept, called "public produce," starts from the belief that food is a human right and therefore the state is obligated to make sure everyone has access to it. An urban designer named Darrin Nordahl coined the term in a book he wrote on the subject when he was working for the city of Davenport in Iowa; he started a small garden in that city where anyone was invited to pick corn, tomatoes, herbs, leafy greens, and more for free, any time. There are many permutations of this same idea. In front of the city hall in Baltimore, instead of the usual showy (and inedible) annuals, leafy greens such as collards and chard were grown in a vegetable garden and then harvested by volunteers to be sent to a local food bank. Vegetables grow in old flowerbeds in front of the Capitol in Madison, Wisconsin, and the food harvested has similarly been donated to a social service agency. In Toronto, a public orchard was created in a midtown park where people in the community care for apricot, apple, plum, and cherry trees and share the harvest. In Seattle, in 2010, enough food was grown on parkland through the city's P-Patch Community Gardens program to provide almost forty-two thousand servings of fresh produce to non-profit agencies. And a parcel of land owned by Seattle Public Utilities is being transformed, with municipal grant money, into the Beacon Food Forest, where people will be free to forage for berries, nuts, and fruit.

In some cases, public produce has been born out of necessity. After the city of Provo, Utah, cut its budget and decided not to put flowers in the planters at city hall, a few municipal planners sowed vegetables instead. They borrowed tools and compost from the city and on their own time tended to their planter gardens and donated the produce to the local food bank. The city of Kamloops, in British Columbia, has fully realized Nordahl's vision of public gardens where people can pick food for free. It has two public produce gardens tended by volunteers where anyone is free to harvest whatever food they like, in addition to a third garden in front of city hall with a similar open-picking policy. "I thought this would be a liberal-versus-conservative issue, but everyone supports this," Nordahl told me. "The liberals think of it as social welfare and the conservatives love it because it is a bit of agrarian tradition. We are teaching some measure of self-sustenance." Some have even turned growing food on public property into revenue-generating businesses. You can buy honey collected from the beehives on the roofs of Chicago's city hall and cultural centre, and on the campus of University of California, Davis, fallen olives that once caused bike accidents are now collected and pressed for their oil, the money earned used to support an olive oil research institution.

On the whole, the gardens have been well received. While some have protested the loss of parkland to vegetable beds, generally people have supported the idea, including politicians. (Though the Baltimore mayor who inherited the vegetable beds on the front lawn reportedly scaled them back because, some speculated, she preferred flowers.) And if the gardens or public orchards are picked clean?

Those who plant them say that demonstrates the need for more free access to fresh fruits and vegetables.

The benefits of urban agriculture are well documented. When someone gardens, they have better access to fresh vegetables, and for this reason, urban agriculture is promoted particularly in lower-income neighbourhoods. In Montreal, the city's network of community gardens makes a real difference in providing fresh produce to people who need it. For example, a quarter of the food grown in the gardens run by the group Action Communiterre is donated to social service organizations, and in another neighbourhood, the average family income of more than half of the people who worked in the community garden was under $20,000, meaning the food they grow helps with the family budget.[49] Gardening also helps people to connect with their neighbours and make friends across community divides. As a number of the academic studies looking at the impacts of urban agriculture have reported, the gardeners noted that once they grew food, their consumption habits changed. They were more likely to support local farmers and to buy consumer products that they believed to be less damaging to the environment.

Something deeper takes place too. "If you asked me why urban agriculture, at the bottom of my list would be growing food," said Charles Levkoe, a PhD candidate at the University of Toronto and a community activist who works with, and writes about, urban agriculture and sustainable food movements. "The real value of people being in the garden is, first of all, to reconnect with one another and regain and relearn the ability to participate together in neighbourhood politics or in municipal politics." Levkoe has witnessed the way gardening vegetables in a city awakens people to civic life, to the potential we all have to make decisions and shape our environment. "Secondly, from the nature perspective," he said, "the value is that nature isn't something over there. It breaks down the artificial separation of society and nature being two different things. We see that we exist as part of that natural system." When we put these two qualities together—a connectedness with nature and the realization that we can have democratic control over our environment—there is the potential for real change. There is the potential for what Levkoe calls a "transformative politics of food."

The philosopher E. O. Wilson argues that we need to have daily contact with nature in order to be happy and healthy human beings. We must connect with nature because our brains have evolved this way. For hundreds of thousands of years, humans constantly interacted with the natural world to sustain themselves, first as hunter-gatherers and then as farmers. Wilson calls this biophilia. It's only been a handful of generations since we've shifted away from this intimate relationship. Because of our cultural distance from biophilia in the modern world, eating is one of the few opportunities we have to make a connection with the earth. So it is through food that we can begin to regain it. Timothy Beatley, who is a professor in the Department of Urban and Environmental Planning at the University of Virginia, has written a book inspired by Wilson titled *Biophilic Cities.* In it, he argues that, seeing as most of humanity lives in an urban environment, we need to reimagine cities as places where we can live with nature and care for it. He builds on Wilson's ideas when he uses the term *biophilic cities.* "In biophilic cities, residents care about nature and work on its behalf, locally and globally," he writes. Urban agriculture makes this idea tangible to many people.

It might seem simplistic to argue that planting kale in a city garden will transform the food system. But we have to start somewhere. And in the garden, we learn about how the worms eat organic matter such as fallen tree leaves and turn the leaves into soil, along with our discarded carrot peelings and apple cores—what we recently thought of as garbage. In the garden we witness the magic of a seed's first tender shoot, the cotyledon pushing out of its casing. We watch as the shoot becomes a seedling, becomes a plant, becomes a flower, becomes a vegetable to eat. When we eat this vegetable, we are familiar with all the steps it took to bring it to our plate, and when we compost the discarded leaves or stems, we know the cycle will start again. We have a taste of biophilia.

Also, as Levkoe described, we learn about democracy in this vegetable bed, particularly if it is in a community garden. We realize we can make decisions about how the garden is run, decide what we will plant and how we will keep this plot of land. This tiny piece of the natural world depends on us—and we start to depend on it. We can decide whether or not it produces food and then we wait to see if nature—the weather, pests, disease—allows this to happen. This feeling of control and at the same time dependency is a powerful experience, even if it is only about a square metre of city soil. It can be the beginning of a cultural awakening, a shift in the way we think about food and the natural world around us. And all the world depends on this shift taking place.

The faster we move now to make this shift happen, the quicker we can build something that lessens our environmental impact, reduces our greenhouse gas emissions, and helps us to adapt to the new climate. The faster we make the shift happen, the faster we can create a new, sustainable way to grow, to produce, to sell, to interact with our food. This is an urgent mission, this unspooling of the industrial food system by 2050, but it is beginning to come undone.

CONCLUSION

Target 2050: The Future

To research this book, I travelled around the world. And when I met people wherever I went, I would be overcome by the strange feeling of having met them before. It didn't matter where they lived, what language they spoke, or how different their lives were from mine, they were familiar to me. That's because they all shared a similar attitude. They had an interest in food as something more profound than only sustenance. They believed that food is culture, that food makes us who we are. And that our relationship to what we eat should be one of balance with one another and with the natural world.

When I was in India, I took a day off from researching to travel to the Ajanta archaeological site, where you can walk into more than two dozen ancient Buddhist caves that were carved into an enormous rock face almost two thousand years ago. The guidebooks suggest visiting the site with a guide, and mine for the day was a thirtyish man from Aurangabad named Sanjay. He was tall and thin and very chatty and before long he was telling me about how it was becoming harder and harder to find what he called desi eggs—desi being slang for all things subcontinental. Whereas he could go to the store and find what were called English eggs, the eggs from new breeds of chickens raised on industrial farms, it was becoming more difficult to buy the desi eggs laid by heritage breeds of chickens. "These are eggs raised by a farmer where the chickens run around here and there, eating what they want," Sanjay explained to me. "The yolks are yellow and they taste better." He lamented their slow disappearance. He might as well have been commiserating with someone I interviewed in my own city who keeps chickens in her backyard, even though it's against the law. She risks keeping hens in her yard in downtown Toronto because she too yearns for eggs laid by real chickens that, just as Sanjay said, run around here and there and eat whatever they want.

In Beijing, I met with a group of students at one of the top universities. A young man who was wearing a yellow and blue windbreaker and sitting with his girlfriend put up his hand. He explained that he was a graduate student in engineering, but really he wanted to be a farmer. He wanted to know if I had any advice for how he could make this transition. The way he spoke reminded me of Monsieur Valadier's son in the Aubrac who, while milking the cows one evening, explained why he wanted to spend his life in the country, as a farmer, rather than live in the city.

Taken on their own, each of these conversations isn't much more than idle small talk. However, put them together—and the many other similar conversations I had—and you've got a thread that connects Sanjay in Aurangabad to the backyard chicken keeper in Toronto to the student in Beijing to the farmer in the Aubrac.

Of course, when a journalist goes looking for stories about the rise of sustainable food systems, it is not a surprise when she finds people interested in creating them. But my random encounters with strangers on three continents does speak to something larger. That is, that there is a universal desire for good food and a concern that the food we want to eat is becoming harder and harder to find. The organization Slow Food, now a global movement, was founded in

Italy in the late 1980s in response to the rise of fast food and what they call the fast life. They have coined a slogan that captures succinctly the qualities that so many of us yearn for in our food: good, clean, and fair. By good food they mean food that is created with care from healthy plants and animals and that expresses our culinary diversity. By clean food they mean food that is good for the health of our bodies as well as the health of the planet. And when they say food that is fair, they mean food that is produced by people who are paid a fair wage for their work and that is accessible and affordable to every human.

But translating these lofty principles into a food system is complicated. There is no straightforward answer because when it comes to food, every rule has many exceptions. For example, on the one hand it is clear that local and sustainable food systems are an economic boon that also have positive social and environmental effects, as opposed to the fallout of industrial food and the corporate-dominated global food economy. But that doesn't mean that in a sustainable food system international trade is verboten. In some cases, it is precisely international markets that help to support the kinds of farming practices and the prosperity we need if we are to feed the world by 2050. Also, in some places, fertilizers derived from fossil fuels might help a lot, whereas in many other cases, this kind of costly external input causes environmental and social harm. Perhaps the heritage seeds of our crops and their wild ancestors hold the answer to sustainable farming in the time of climate change. But maybe publicly funded, science-based seed breeding, conducted in a patent-free, corporate-free system holds the key to future food security.

The future of food is not sound bite friendly.

Food might be complicated, but it is possible to distill a basic set of criteria we can use to measure a future system and guide us towards creating something better than what we have today—a good, clean, and fair food system.

To start, any food system must be sustainable. By that I mean it must be able to not only provide us with food today but continue to provide food well into the future. Our farming can't take more from the biosphere than it gives back. Our food system must be in balance. What that looks like in practice are farms where the soil is healthy, where the nutrients are managed responsibly, and where close attention is paid to the micro-organisms that live below the surface. On these farms, water should also be stewarded with utmost concern both for conservation and for effluent—we must not pollute the water with fertilizer runoff or with livestock waste. All this must be done while paying attention to energy use; we must adopt fossil-fuel-free energy sources now.

Secondly, it is important that our world's farmers earn a living wage. We cannot continue to exploit farmers so the rest of us can eat cheap food. Because the big food corporations mostly have helped to further marginalize these people, we have to build food systems that circumvent the corporate middlemen. While this will inevitably mean that we must strengthen local food economies, we have to ensure that any trade in food supports the basic values of good, clean, fair food. Exploitation in international trade should be prevented with fair trade certification that is able to lessen the chance that farmers in faraway places—and their environments—will suffer. And transit needs to be improved too. We need to ship products from one place to another in the most carbon neutral way possible.

In this book, I haven't documented how our food system abuses animals. The treatment of animals in the global economy of food is frequently cruel and inhumane. We must treat the livestock we eat with the respect they deserve. To start, that will mean an end to industrial livestock operations and a return to pasture-based farming systems. Yes, this will mean a dramatic reduction in the meat supply. But this is good, not only for our health (we eat too much of the stuff anyway) but also for the world—the meat industry is responsible for 18 percent of our greenhouse gas emissions. Then we must continue to strive to improve the way we treat these living beings.

To support sustainable food systems everywhere, we must protect farmland for posterity. We must preserve the land that grows our food around the cities and towns and stop turning farmers' fields into quarries and mines and vast biofuel plantations. Rather than expanding food production on the soil in someone else's backyard, as we are doing now in the global rush for farmland, we should focus on sustainable intensification, trying to extract more yields from the land that we already farm. This will require funding more research into improving sustainable farming. And rather than focusing only on technological fixes, we need public funding for old-fashioned agricultural extension to breed new seeds and new livestock breeds as well as develop the most environmentally positive ways to grow our food and our textiles too.

At base, we need a democratic system. We must adopt the idea of food sovereignty as advocated by La Via Campesina so we can all choose what food we eat and how this food is produced. And to do this, we must work together.

In his book *Here on Earth,* the internationally acclaimed Australian scientist and conservationist Tim Flannery puts forward the idea that a human superorganism has the potential to save both the human species from an uncertain future and the ecological systems on planet Earth from massive collapse. Despite the gruesome environmental truths that we face in what he calls "this century of decision," he is optimistic about our ability to work together. He writes that we must collaborate as a species, across national, cultural, political, and linguistic barriers to manage the global commons—the atmosphere and the oceans that we all depend on and share. We must work together as a global superorganism.

His idea of a human superorganism was inspired by ants. The fire ants of North America act differently from other ant colonies. Rather than marking and protecting its colonies as most species of ants do, this population of fire ants has a genetic difference that allows it to view the ant colony next door as being friendly rather than threatening. They see their neighbours as being part of the same larger community, the same superorganism. Humans, Flannery posits, are the global version of such a superorganism. Human migration, intermarriage between cultures and races, transnational trade in food and other goods, the Internet—all this globalization has brought us closer together than we have ever been before, completely interconnecting us as a species. The global superorganism has been formed, he announces. To save ourselves and our planet, we must be humble, he writes. We must recognize our dependency on the ecosystems that we are part of. We must love the planet and our fellow humans as much as we love ourselves.

There are signs of Flannery's global superorganism in all the manifestations of the global social movement that is working towards creating food systems that are fair and just and that can achieve an ecological balance. All these grassroots efforts scattered across the continents in thousands and thousands of communities are the twitching nerves of the superorganism as it takes shape and comes to life. That Kenyans are trying to rebuild sustainable food systems and a holistic food culture in the same way as Americans and Canadians and Indians and Ethiopians and Greeks and Japanese and Koreans and Thais is no coincidence. Quite the opposite. It is of enormous significance.

When I was in Beijing, I met a university professor who had helped a small rice-growing village shift from depending on the wages of those who had left to work in the factories to being able to grow rice and sell it for a good price directly to people in the city. "How can a farmer live from rice? This is my wish," she told me. But then she sighed and hunched her shoulders, looked downcast. All this work she had done to help this one village, she said, was, in the grand scheme, "very small. It doesn't change life fundamentally. If you want to make more money, you let your husband and your son go to the city to do non-agricultural work."

True, if you see each tiny effort in isolation, it does appear to be small, even insignificant in the face of the behemoth that is industrial food. But taken together, these small efforts equal profound change. This isn't about only one village in China. If you were to take a map of the globe and indicate all these tiny projects—and the larger ones too—with little red dots, a form would start to appear. Connect these dots and the outline of the superorganism would take shape.

The food system is an ideal vehicle for Flannery's superorganism because it involves everyone. We all must eat. Flannery says we need to love one another just as much as we love ourselves. What better way to come together as communities than through the breaking of bread? He says we need to love the planet, too, as much as we love ourselves. Again, food offers us the opportunity to connect with nature, to realize that we are a part of the cycles of life on this planet, and that by loving nature, we are loving ourselves. Through building good, clean, and fair food systems in communities around the world, we can find this opportunity to love one another, to love this finite planet and move into the future together.

ACKNOWLEDGEMENTS

When my first book, *Locavore,* was published, I was launched into a growing debate about the future of food and I had the impression that the other side was winning in the theatre of public opinion. At speaking engagements, at dinner parties, in off-hand conversations, someone would make the point that, although sustainable food systems sounded like the right course for society, we had to be realistic about the future and abandon the right for the might of industrial food. I knew this point of view was based on incomplete information, and so I decided to delve further into the debate to provide a more complete picture of the phenomenal benefits of sustainable food systems. So I spent two years travelling, interviewing, and assembling a response to all these realists: we can be realistic *and* optimistic at the same time. Right feels right because it is right.

I would like to thank the people who provided me with the fodder for this book. There are so many. First of all, I want to thank three people who allowed me to turn them into the three main characters of my book. Chandrakala Bobade in Dhangaon, India, shared so willingly her life story with a woman from Canada. Rose in China helped out in so many ways. She acted as guide and translator and was even able to move boulders—really—when a landslide in the mountains blocked the road. But it was Rose's nostalgic descriptions of her life growing up among the paddies that moved me so much. And thank you to André Valadier in the Aubrac. I developed a deep admiration for this incredible man and I believe that by the end of my stay, we both felt as though we'd known each other for much longer. His energy is so intense that it still buoys me!

Thank you also to Nikki Eason in Detroit, Mario Duchesne of the Association for the Development of the Canadienne Cattle Breed, and to Shi Yan and her husband at Little Donkey Farm for hosting a terrific lunch and introducing me to the world of human manure. Thanks to Bryan and Cathy Gilvesy for sharing their wild enthusiasm for farming, food, and the environment.

I am greatly indebted to Dr. Tony Fuller, professor emeritus at the University of Guelph, who made my trip to China possible and who shared with me all of his own research and contacts and also contributed many laughs to the journey. A tremendous thank you to Dr. Tammy Sage in Toronto for her aid in navigating the science behind her work. I am grateful to Gauri Mirashi, who translated for me in Aurangabad and area, and to Joy Daniel as well as Navnath Dhakane of the Institute for Integrated Rural Development, who helped me get the most out of my research trip. Raghu Rao in Bangalore made sure to remove any rose-coloured glasses and took time to show me what he felt to be the real story of sustainable farming in India—thanks to him as well.

I'd like to thank Dr. Evan Fraser at the University of Guelph for reading an earlier version of the manuscript and sharing his amazing expertise. Dr. Sarah Gower is not only a dear friend but a smart reader, and I thank her for her help. Thanks to my cousin Dr. Justin Robertson at the University of Hong Kong for his insight, as well as to *mashi* Piali Roy and my sister Elyssa Elton, who endured endless questions. Also, thank you to Jo-Ann and Hugh Robertson. A huge thanks to my friend Sarmishta Subramanian too.

I didn't expect that writing a book would lead to so many new friendships, but now, thanks to books, I count many more friends around the world. Thank you to everyone—too numerous to name—for being so open and warm and curious.

At HarperCollins Canada, a big thank you to my editor, Jim Gifford, for believing in, and editing, this book, and to the unstoppable Rob Firing for everything that he does professionally and also for his contagious passion for good, clean, fair food. Thank you to Shaun Oakey for his astoundingly thorough copy edit and impressive knowledge base. I would also like to acknowledge the role my agent, Samantha Haywood of the Transatlantic Literary Agency, played in this book. Not only does she represent my interests but she has been a sounding board for ideas and an all-round special adviser. Thank you.

Thank you also to Ameena Sultan, Shahnaz Khan, Bruce Roberts, and Bruce Walsh for ideas, discussion, and friendship. Thank you to my family: Elyssa Elton (this time for being my sister) and Paul Gorman; Amtu and Nurdin Karimjee; a tremendous thank you to my parents, Peter and Jacqueline Elton, for travel companionship and babysitting and so much more; to Nadia and Anisa for being the best reason in the world to leave my desk. I'd also like to thank my grandmother, Marjorie Abrams, who might not understand what I've been up to for these past five years but whose imprint on my life has been indelible and who has undoubtedly shaped this book.

Finally, my love and gratitude for Kumail is unending. Without your tremendous support, none of this would ever have been possible.

Let us all, as the old folks say in China, find *jie di qi.*

A note on translation: A good part of the research for this book was conducted in French, including interviews in the field. I would like to thank my mother, Jacqueline Elton, a career translator, for helping to ensure the accuracy of language in these quotes as I translated them to English.

NOTES

Chapter 1: Table for One Billion

1. Upali A. Amarasinghe, Tushaar Shah, and B. K. Anand, "India's Water Supply and Demand from 2025–2050: Business-as-Usual Scenario and Issues" (New Delhi: International Water Management Institute, 2008), http://www.iwmi.cgiar.org/publications/Other/PDF/NRLP Proceeding-2 Paper 2.pdf (accessed January 2012).

2. Shabab Fazal, "Urban Expansion and Loss of Agricultural Land," *Environment and Urbanization* 12, no. 2 (2000): 133.

3. Atiqur Rahman and Maik Netzband, "An Assessment of Urban Environmental Issues Using Remote Sensing and GIS Techniques: An Integrated Approach. A Case Study: Delhi, India" (2009), http://www.ciesin.columbia.edu/repository/pern/papers/urban_pde_rahman_etal.pdf (accessed January 2012).

4. P. K. Aggarwal, "Vulnerability of Indian Agriculture to Climate Change: Current State of Knowledge" (New Delhi: Indian Agricultural Research Institute, October 2009), http://moef.nic.in/downloads/others/Vulnerability_PK Aggarwal.pdf (accessed November 2012).

5. William R. Cline, "Global Warming and Agriculture," *Finance and Development* 45, no. 1 (March 2008), http://www.imf.org/external/pubs/ft/fandd/2008/03/cline.htm (accessed December 2012).

6. Praduman Kumar, "Cereals Prospects in India to 2020," International Food Policy Research Institute Brief, 1995, http://www.ifpri.org/sites/default/files/publications/vb23.pdf (accessed January 2012).

7. US Geological Survey, "Groundwater Depletion," http://ga.water.usgs.gov/edu/gwdepletion.html (accessed December 2012).

Chapter 2: Faster, Bigger, Richer, Weaker

8. Jules Pretty, *Agri-Culture: Reconnecting People, Land and Nature* (London: Earthscan, 2002), 4.

9. J. R. McNeill, *Something New Under the Sun: An Environmental History of the Twentieth-Century World* (New York: W. W. Norton, 2001), 222.

10. H. Charles J. Godfray, "Food and Biodiversity," *Science* 333, no. 6047 (2011): 1231.

11. Eric Holt-Giménez, "Measuring Farms Agroecological Resistance to Hurricane Mitch," *Agriculture, Ecosystems and Environment* 93 (December 2002): 87.

12. Peter Rosset, "Small Is Bountiful," *The Ecologist* 29 (December 1999); Miguel A. Altieri, Fernando R. Funes-Monzote, and Paulo Petersen, "Agroecologically Efficient Agricultural Systems for Smallholder Farmers: Contributions to Food Sovereignty," *Agronomy for Sustainable Development* (November 2011), http://www.agroeco.org/socla/pdfs/Altieri-Funes-Petersen-Palencia.pdf (accessed February 2012).

13. Andrew Dorward, "Farm Size and Productivity in Malawian Smallholder Agriculture," *Journal of Development Studies* 35 (1999): 141.

14. Altieri et al., "Agroecologically Efficient Agricultural Systems."

15. J. N. Pretty, A. D. Noble, D. Bossio, J. Dixon, R. E. Hine, F. W. T. Penning de Vries, and J. I. L. Morison, "Resource-Conserving Agriculture Increases Yields in Developing Countries," *Environmental Science and Technology* 40, no. 4 (2006): 1114.

Chapter 3: The Money Knot

16. Olivier De Schutter, "The World Trade Organization and the Post-Global Food Crisis Agenda: Putting Food Security First in the International Trade System" (briefing note by the special rapporteur on the right to food, Louvain, November 2011), 3.

17. National Farmers Union, "Farmers, the Food Chain and Agriculture Policies in Canada in Relation to the Right to Food" (submission to the special rapporteur on the right to food, May 2012).

18. Michael Carolan, *The Real Cost of Cheap Food* (New York: Earthscan, 2011), 196.

19. E. Wesley F. Peterson, *A Billion Dollars a Day: The Economics and Politics of Agricultural Subsidies* (Malden, MA: Wiley-Blackwell, 2009), 10.

20. Peterson, *A Billion Dollars a Day,* 10.

21. Peterson, *A Billion Dollars a Day,* 141.

22. De Schutter, "The World Trade Organization and the Post-Global Food Crisis Agenda," 13.

Chapter 4: Local versus Industrial

23. "The Economics of Urban Agriculture," editorial, RUAF *Urban Agriculture Magazine,* no. 7 (2002).

Chapter 5: The Twenty-First-Century Peasant

24. Nadia El-Hage Scialabba, "Organic Agriculture and Food Security" (paper presented at International Conference on Organic Agriculture and Food Security, Italy, May 2007), 13.

25. Altieri et al., "Agroecologically Efficient Agricultural Systems."

26. Maria Elena Martinez-Torres and Peter M. Rosset, "La Via Campesina: The Birth and Evolution of a Transnational Social Movement," *Journal of Peasant Studies* 37, no. 1 (2010): 165.

Chapter 6: Land as Good as Gold

27. Lorenzo Cotula, Sonja Vermeulen, Rebeca Leonard, and James Keeley, "Land Grab or Development Opportunity? Agricultural Investment and International Land Deals in Africa" (Rome and London: IIED/FAO/IFAD, 2009), http://www.ifad.org/pub/land/land_grab.pdf (accessed December 2012).

28. Oxfam, "Land and Power: The Growing Scandal Surrounding the New Wave of Investments in Land" (Oxfam Briefing Paper 151, September 2011), http://www.oxfam.org/sites/www.oxfam.org/files/bp151-land-power-rights-acquisitions-220911-en.pdf (accessed December 2012).

29. "Land Grabbing and the Global Food Crisis" (presentation, GRAIN, December 16, 2011), http://www.grain.org/article/entries/4164-land-grabbing-and-the-global-food-crisis-presentation (accessed January 2013).

30. Chengri Ding, "Farmland Preservation in China," *Land Lines* 16, no. 3 (Lincoln Institute of Land Policy, July 2004), http://www.lincolninst.edu/pubs/913_Farmland-Preservation-in-China (accessed December 2012).

31. Joëlle Noreau with Jean-Michel Goulet, "Foreign Purchase of Québec Farmland: A Takeover or a Misunderstanding?" (Desjardins Economic Studies, *Economic Viewpoint,* May 11, 2010), 36, http://www.desjardins.com/en/a_propos/etudes_economiques/actualites/point_vue_economique/pv0511a.pdf (accessed December 2012).

Chapter 7: Two Thousand Years of Rice

32. Na Guo, "China's Mountain Regions: How to Maintain an Environmentally Sustainable Development While Improving Livelihoods? A Case Study of Yuanyang County, Yunnan Province, Southwest Part of China" (master's thesis, Lund University, 2010), http://www.lumes.lu.se/database/alumni/08.10/Thesis/Guo_Na_Thesis_2010.pdf (accessed December 2010).

33. C. Picone and D. L. Van Tassell, "Agriculture and Biodiversity Loss: Industrial Agriculture," in *Life on Earth: An Encyclopedia of Biodiversity, Ecology, and Evolution,* ed. Niles Eldredge (Santa Barbara: ABC-CLIO, 2002), http://www.landinstitute.org/pages/Picone and Van Tassel 2002.pdf (accessed December 2012).

Chapter 8: The Genes in Our Seeds

34. Philip H. Howard, "Visualizing Consolidation in the Global Seed Industry: 1996–2008," *Sustainability* 1, no. 4 (2009): 1269.

35. Seed Savers Exchange, *Garden Seed Inventory: An Inventory of Seed Catalogs Listing All Non-Hybrid Seeds Available in the United States and Canada,* 6th ed. (Decorah, IA: 2006), 11.

Chapter 9: Lab Rice

36. Angelo Katsoras and Pierre Fournier, "What the Looming Food Crisis Means for Investors" (National Bank Financial, June 2009), 15.

37. Center for Food Safety, "Monsanto vs. U.S. Farmers" (Washington, DC: Center for Food Safety, 2005), http://www.centerforfoodsafety.org/pubs/CFSMOnsantovsFarmerReport1.13.05.pdf (accessed December 2012).

38. Doug Gurian-Sherman, "Failure to Yield: Evaluating the Performance of Genetically Engineered Crops" (Union of Concerned Scientists, 2009), http://www.ucsusa.org/food_and_agriculture/our-failing-food-system/genetic-engineering/failure-to-yield.html (accessed December 2012).

39. "Monsanto Cited in Crop Losses," *New York Times,* June 16, 1998, http://www.nytimes.com/1998/06/16/business/monsanto-cited-in-crop-losses.html?src=pm (accessed December 2012).

Chapter 10: SOS

40. Nick Cullather, "Miracles of Modernization: The Green Revolution and the Apotheosis of Technology," *Diplomatic History* 28, no. 2 (April 2004): 243.

Chapter 11: From Home-Cooked to Takeout

41. US Energy Information Association, "Cooking Trends in the United States: Are We Really Becoming a Fast Food Country?" http://www.eia.gov/emeu/recs/cooking-trends/cooking.html (accessed December 2012).

42. Alberta, Ministry of Agriculture and Rural Development, "Understanding Consumer Trends Can Present New Opportunities," http://www1.agric.gov.ab.ca/$department/deptdocs.nsf/all/sis8735 (accessed December 2012).

43. Vani S. Kulkarni and Raghav Gaiha, "Dietary Transition in India" (Philadelphia: Center for the Advanced Study of India, 2010), http://casi.ssc.upenn.edu/iit/kulkarni-gaiha (accessed December 2012).

44. Oxfam International, "GROW Campaign 2011: Global Opinion Research—Final Topline Report" (May 2011), http://www.oxfam.org/sites/www.oxfam.org/files/grow-campaign-globescan-research-presentation.pdf (accessed November 2012).

Chapter 12: The Terroirists to the Rescue!

45. Maud Hirczak, Mehdi Moalla, Amédée Mollard, Bernard Pecqueur, Mbolatiana Rambonilaza, and Dominique Vollet, "Le modèle du panier de biens," *Économie Rurale* no. 308 (November-December 2008), http://economierurale.revues.org/366 (accessed August 2011).

46. Amy B. Trubek, *Haute Cuisine: How the French Invented the Culinary Profession* (Philadelphia: University of Pennsylvania Press, 2000), 35.

47. Parviz Koohafkan and Miguel Altieri, *Globally Important Agricultural Heritage Systems,* http://www.fao.org/fileadmin/templates/giahs/PDF/GIAHS_Booklet_EN_WEB2011.pdf (accessed September 2011).

Chapter 14: Introducing . . . Food

48. Monica Caggiano, "Les 'Jardins Partagés' in Paris: Cultivating Visions and Symbols" (paper presented at 9th European IFSA Symposium, Vienna, July 2010), 1216.

49. E. Duchemin, F. Wegmuller, and A.-M. Legault, "Urban Agriculture: Multi-dimensional Tools for Social Development in Poor Neighbourhoods," *The Journal of Field Actions, Field Actions Science Reports 1* (2008), http://factsreports.revues.org/113 (accessed December 2012).

SELECTED BIBLIOGRAPHY

Altieri, Miguel A. *Agroecology: The Scientific Basis of Alternative Agriculture.* Boulder: Westview Press, 1987.

Altieri, Miguel A., and Parviz Koohafkan. *Enduring Farms: Climate Change, Smallholders and Traditional Farming Communities.* Malaysia: Third World Network, 2008.

Andrews, Neil, David Bailey, and Ivan Roberts. *Agriculture in the Doha Round.* London: Commonwealth Secretariat, 2004.

Beatley, Timothy. *Biophilic Cities: Integrating Nature into Urban Design and Planning.* Washington, DC: Island Press, 2011.

Bloom, Jonathan. *American Wasteland: How America Throws Away Nearly Half of Its Food.* Cambridge, MA: Da Capo Press, 2010.

Carolan, Michael. *The Real Cost of Cheap Food.* New York: Earthscan, 2011.

Chen, Guidi, and Wu Chuntao. *Will the Boat Sink the Water? The Life of China's Peasants.* Translated by Zhu Hong. New York: Public Affairs, 2006.

Cockrall-King, Jennifer. *Food and the City.* New York: Prometheus, 2012.

Commission on Genetic Resources for Food and Agriculture. "The Second Report on the State of the World's Plant Genetic Resources for Food and Agriculture." Rome: Food and Agriculture Organization, 2010.

Daniel, Charles. *Lords of the Harvest: Biotech, Big Money, and the Future of Food.* Cambridge, MA: Perseus Books, 2001.

Cullather, Nick. *Hungry World: America's Cold War Battle Against Poverty in Asia.* Cambridge, MA: Harvard University Press, 2010.

Diamond, Jared. *Collapse: How Societies Choose to Fail or Succeed.* New York: Viking, 2005.

Fernández-Armesto, Felipe. *Near a Thousand Tables: A History of Food.* New York: Free Press, 2004.

Flannery, Tim. *Here on Earth: A Natural History of the Planet.* Toronto: HarperCollins Canada, 2010.

Fraser, Evan D. G., and Andrew Rimas. *Empires of Food: Feast, Famine and the Rise and Fall of Civilizations.* New York: Free Press, 2010.

Garner, Carley. *A Trader's First Book on Commodities: An Introduction to the World's Fastest Growing Market.* New Jersey: FT Press, 2010.

George, Rose. *The Big Necessity: The Unmentionable World of Human Waste and Why It Matters.* New York: Henry Holt, 2008.

Gorgolewski, Mark, June Komisar, and Joe Nasr. *Carrot City: Creating Places for Urban Agriculture.* New York: The Monacelli Press, 2011.

Kingsbury, Noel. *Hybrid: The History and Science of Plant Breeding.* Chicago: University of Chicago Press, 2009.

Kloppenburg, Jack. *Seed Sovereignty: Reconnecting Food, Nature and Community.* Winnipeg and Black Point, NS: Fernwood, 2010.

Koohafkan, Parviz, and Miguel A. Altieri. *Globally Important Agricultural Heritage Systems: A Legacy for the Future.* Rome: Food and Agriculture Organization, 2011.

Kuyek, Devlin. *Good Crop/Bad Crop: Seed Politics and the Future of Food in Canada.* Toronto: Between the Lines, 2007.

Levine, Susan. *School Lunch Politics: The Surprising History of America's Favorite Welfare Program.* New Jersey: Princeton University Press, 2008.

McNeill, J. R. *Something New Under the Sun: An Environmental History of the Twentieth-Century World.* New York: W. W. Norton, 2001.

Metcalf, Barbara D., and Thomas R. Metcalf. *A Concise History of Modern India.* Cambridge: Cambridge University Press, 2006.

Moore Lappé, Frances, and Anna Lappé. *Hope's Edge: The Next Diet for a Small Planet.* New York: Penguin, 2003.

Nabhan, Gary Paul. *Where Our Food Comes From: Retracing Nikolay Vavilov's Quest to End Famine.* Washington, DC: Island Press, 2009.

———. *Why Some Like It Hot: Food, Genes, and Cultural Diversity.* Washington, DC: Island Press, 2004.

Newman, Jonathan A., Madhur Anand, Hugh Henry, Shelly Hunt, and Ze'ev Gedalof. *Climate Change Biology.* Oxford: CABI, 2011.

Nordahl, Darrin. *Public Produce: The New Urban Agriculture.* Washington, DC: Island Press, 2009.

Nützenadel, Alexander, and Frank Trentmann, eds. *Food and Globalization: Consumption, Markets and Politics in the Modern World.* Oxford: Berg, 2008.

Patel, Raj. *Stuffed and Starved: Markets, Power and the Hidden Battle for the World's Food System.* Toronto: HarperCollins Canada, 2007.

Perfecto, Ivette, John Vandermeer, and Angus Wright. *Nature's Matrix: Linking Agriculture, Conservation and Food Sovereignty.* London: Earthscan, 2009.

Peterson, E. Wesley F. *A Billion Dollars a Day: The Economics and Politics of Agricultural Subsidies.* Malden, MA: Wiley-Blackwell, 2009.

Pistorious, Robin. *Scientists, Plants and Politics: A History of the Plant Genetic Resources Movement.* Rome: International Plant Genetic Resources Institute, 1997.

Pollan, Michael. *The Omnivore's Dilemma: A Natural History of Four Meals.* New York: Penguin, 2006.

Pretty, Jules. *Agri-Culture: Reconnecting People, Land and Nature.* London: Earthscan, 2002.

———. *Regenerating Agriculture: Policies and Practice for Sustainability and Self-Reliance.* Washington, DC: Joseph Henry Press, 1995.

Raymond, Hélène, and Jacques Mathé. *Une agriculture qui goûte autrement: Histoires de productions locales de l'Amérique du nord à l'Europe.* Quebec: Éditions MultiMondes, 2011.

Ronald, Pamela C., and Raoul W. Adamchak. *Tomorrow's Table: Organic Farming, Genetics, and the Future of Food.* Oxford: Oxford University Press, 2008.

Sen, Amartya. *Development as Freedom.* New York: Knopf, 1999.

Simmonds, N. W., ed. *Evolution of Crop Plants.* London: Longman, 1976.

Smil, Vaclav. *Enriching the Earth: Fritz Haber, Carl Bosch, and the Transformation of World Food Production.* Cambridge, MA: MIT Press, 2001.

Trubek, Amy B. *The Taste of Place: A Cultural Journey into Terroir.* Berkeley: University of California Press, 2008.

Van der Ploeg, Jan Douwe. *The New Peasantries: Struggles for Autonomy and Sustainability in an Era of Empire and Globalization.* London: Earthscan, 2009.

Vavilov, N. I. *Five Continents.* Translated by Doris Love. Rome: International Plant Genetic Resources Institute, 1997.

Weis, Tony. *The Global Food Economy: The Battle for the Future of Farming.* Winnipeg and Black Point, NS: Fernwood, 2007.

Wilson, Edward O. *Biophilia.* Cambridge, MA: Harvard University Press, 1984.

Winders, Bill. *The Politics of Food Supply: U.S. Agricultural Policy in the World Economy.* New Haven: Yale University Press, 2009.

Worldwatch Institute. *State of the World 2011: Innovations That Nourish the Planet.* New York: Worldwatch, 2010.

Yiching, Song, and Ronnie Vernooy. *Seeds and Synergies: Innovating Rural Development in China.* Ottawa: International Development Research Centre, 2010.

INDEX